Paths, Tracks, and Trails

DESIGNING FOR PEDESTRIANS
AND CYCLISTS

Edited by
Paolo Ceccon & Laura Zampieri

Paths, Tracks, and Trails

DESIGNING FOR PEDESTRIANS AND CYCLISTS

Foreword

Noel Corkery
Corkery Consulting

Noel Corkery is the Managing Director of Corkery Consulting, an award-winning urban design and landscape architecture studio based in Sydney, specialising in planning and design of public spaces and transport infrastructure. Noel has worked on projects throughout Australia as well as Hong Kong and China. He holds a Master of Landscape Architecture degree from Cornell University, as well as an MBA and Master of Cross Disciplinary Art & Design from the University of NSW, and a Bachelor of Science in Forestry from the Australian National University. Noel is a Fellow of the Australian Institute of Landscape Architects and past National President. Noel received the AILA President's Award for his outstanding contribution to the landscape architecture profession in Australia.

The evolution of cities involves a continuous process of development and redevelopment. A key component of this process is the infrastructure that allows the movement of people throughout cities and beyond.

While roads continue to be the dominant form of urban infrastructure in most cities, this is changing. There is an increasing emphasis on public space infrastructure associated with greater urban densities. There is also a growing acknowledgement, backed by research, that walking and cycling can be major contributors to the health and well-being of communities.

The spectrum of public space infrastructure throughout cities encompasses parks, streets, civic plazas, markets, sports facilities, scenic lookouts, riverfronts and coastal foreshores. The physical form of public space infrastructure is not only a functional issue, but strongly influences how people interact with each other in those spaces. Creatively designed public spaces connected by a network of walking and cycling pathways have been shown to encourage and facilitate interaction and collaboration between communities by reducing segregation and division. Consequently, many designers are now focusing on how to create public spaces that facilitate safe, engaging and equitable social interaction.

Urban redevelopment increasingly involves the transformation of former industrial land into living spaces for new communities. The pattern of streets associated with these areas seldom provides for pedestrian and cyclist movement. It therefore is necessary to develop walking and cycling path networks that provide connectivity between communities and public spaces, schools, recreation and cultural facilities, commercial centers and employment locations.

In many older urban areas, existing infrastructure is reaching the end of its economic life. The need for renewal provides opportunities to incorporate new walking and cycling path systems and public spaces as key components of urban infrastructure. This integrated approach to infrastructure provides the means of creating highly connected built and natural environments.

Successful design of these public spaces results in multi-functional walking and cycling corridors and public spaces that are convenient, enjoyable and inclusive.

The complexity of public space infrastructure demands the involvement of multiple disciplines that often includes engineers (structural, civil, hydraulic, traffic, acoustics), environmental and sustainability specialists, architects, planners (strategic, statutory, social), community consultation facilitators, heritage specialists, archaeologists, public artists, horticulturalists and cost planners. Good public spaces are often the result of many hands and minds working creatively together. The design process in which these diverse disciplines are involved therefore requires coordination. Increasingly, landscape architects are providing the leadership required to coordinate multi-disciplinary teams to design new public space infrastructure.

A major challenge for multi-disciplinary design teams is maximizing the ease with which people move along and across streets, railways, rivers and waterfronts, between and through buildings. The density of walking and cycling path networks throughout a city determines the ease of movement, or porosity. There is growing recognition that highly porous cites offer greater opportunities for social engagement, interaction and resilience. In cities that encourage active, personal transport—walking and cycling—there are fewer cars on the streets, lower carbon emissions, and cleaner air. The slower pace of walking and cycling through cities allows people to engage more directly with the landscape and city life.

Connectivity between the places where people live, work and play should not only be functional, but also encompass the poetics of movement through time and space. Good design of public space infrastructure engages the sensory experiences of sight, sound, touch, and smell as well as the kinesthetic awareness of balance and movement. Further, it plays with the way the character and quality of a space changes over the course of a day and from one season to the next. The projects presented in this book demonstrate an impressive international cross-section of creative solutions to the public space infrastructure challenges created by high urban density. The projects will provide inspiration and a benchmark for other designers and communities striving for high-quality public space infrastructure.

However, each project is unique and the result of a design process that responds to its particular physical, social, economic, cultural, and political circumstances. Designers need to understand and respond to the complex combination of these influences.

Challenges to the design of successful public space infrastructure include:
- the relation of public spaces to the overall spatial organization of a city and the movement of people through them
- how the infrastructure physically relates to its context
- people's movement through and their perception of the built and natural environment of cities
- the role of infrastructure as public space.

The projects presented on the following pages illustrate that the issues of connectivity and urban public space are common across different cultural and geographic settings. At the same time, solutions differ in response to the character of individual sites and their context within an urban or natural landscape setting.

They demonstrate a commitment to creative design to produce new public spaces that are not only functional, but provide opportunities for people to move easily through them and interact with the surrounding urban and natural environments. This outcome has been achieved through the creation of paths, tracks, and trails for the use of pedestrians and cyclists.

Contents

4	Foreword
8	Introduction
9	1 Public Infrastructure and Public Spaces
14	2 Planning a Walking and Cycling Path System
18	3 Types and Locations of Walking and Cycling Paths
25	4 Path-User Requirements
29	5 Design Criteria for Walking Paths
36	6 Design Criteria for Bicycle Paths
42	7 Intersections of Paths and Roads
44	8 Provision Parking Facilities
46	9 Maintenance and Management
47	Bibliography

Case Studies

Paths, tracks, and trails for pedestrians
and cyclists on waterfronts

48	Revitalisation of Spikeri Square and Daugava Waterfront Promenade
56	Mapocho 42K Riparian Promenade
64	Renewing the Wharf of Austerlitz Port
72	Krymskaya Embankment
78	Westhaven Promenade
84	La Perouse Headland Coastal Walk and Loop Road
92	Kalvebod Waves

Paths, tracks, and trails for pedestrians and cyclists in urban streets

- 100 Central Dandenong Lonsdale Street redesign and upgrade
- 106 Padua Train Station Square
- 114 Piazza Nember
- 120 Adolf Horn Avenue
- 126 Bourke Street cycleway
- 132 George Street cycleway
- 140 Queens Quay Boulevard
- 146 Abercrombie Street upgrade

Paths, tracks, and trials for pedestrians and cyclists in parks and other green spaces

- 152 Madrid Rio
- 160 Governors Island, Phase 1—Park and Public Space
- 166 New Park for the University Quarter in Essen
- 174 Torreblanca Aromatic Park in Torrevieja
- 180 The Erbe Danzanti Park
- 186 The Porto—Lerone Cycle Lane
- 190 Northwest Arkansas Razorback Regional Greenway
- 198 Te Ara/Alexandra Stream bike path

Bridges and underpasses for pedestrians and cyclists

- 204 The Luchtsingel
- 210 Bicycle Underpass, Haarlem
- 214 El Valle Trenzado
- 220 Glacier Skywalk
- 226 Energieberg Georgswerder Horizontweg
- 234 Hovenring bicycle bridge
- 240 The Paleisbrug
- 248 Vlotwateringbrug
- 254 La Sallaz Footbridge
- 258 Pedestrian Bridge in Aranzadi Park
- 264 The Bicycle Snake

270 Index

Introduction

This book sets out possible ways to improve walking and cycling environments. It outlines a process for helping people to understand what type of infrastructure should be provided for pedestrians and cyclists.

The book collects convergent points of view from different international laws and guidelines regarding walking and cycling environments, and analyzes the most advanced worldwide rules, in particular, social/environmental targets and geometrical/material standards of the European Union (mainly UK, Netherlands, Italy, and Switzerland), United States, New Zealand, and Australia. It transcribes and incorporates extracts—and in some cases, entire concepts—from the main international laws and guidelines, updating and critiquing the contents. In this sense, the references to the geometrical data should not be treated as instructions, but as general tips to be considered before checking with the national legislations for matching the real-world conditions of a place (type of society, national and local policy, economic trends, environmental and geographic context, climate, and so on). The physical elements of pedestrian and cyclist enviroments are discussed in light of the fact that sustainable mobility and ecological infrastructure are the main trends for new urban spaces. The text and the projects discussed in this book also show that walking and cycling systems have to be considered as social components, which cannot be described merely through a separate technical regulation. They should be considered within a broader understanding of the concept of mobility.

The book is oriented to help people involved in planning paths and tracks for pedestrians and cyclists. It encourages the designers who develop urban walking and cycling paths, tracks, trails, and networks to rediscover walking and cycling as viable modes of transportation for short trips in and around neighborhoods and cities. It also highlights the importance of recognizing these paths, track and trails as integrated with the main public transportation systems, and also to consider their tourist potential.

More than a planning and design guide, this book intends to serve as a key to interpreting the following projects' best practices, which highlight the application of the most advanced regulatory developments and design criteria. By providing design advices and standards, it promotes a consistent "world's best practice" approach to planning, designing, operating, and maintaining walking and cycling network infrastructure. In this sense, walking and cycling mostly take place within a transportation system that must work for a range of road users. This requires effectively integrating pedestrian and cyclist access and safety provisions for travelling along and across roads, with routes outside road corridors as part of a continuous network. This guide applies to all "slow mobility" infrastructure, whether it is alongside or across roads, through parks, public gardens, and sports and recreational areas, or on private land where public presence might reasonably be expected. It also applies to new developments, facility changes, and existing environments.

However, the pedestrian and cyclist planning approaches, construction design, and interventions adopted will depend on the circumstances at each location. The financial and technical, as well as contingent, factors or the local regulations and national legislation may affect what can be achieved at any particular location or time. This book is only a general guide that does not prescribe a single approach or intervention, but presents completed projects along with their advantages, disadvantages and limitations, which are useful as planning and design tools.

The guide collects and reports on the main criteria, indicative rather than prescriptive, of the planning and design derived from the combination of guidelines, regulations, and the most advanced design examples.

1 Public Infrastructure and Public Spaces

The re-thinking of the idea of mobility infrastructure linked to sustainability cannot only refer to specific walking and cycling paths, but must also include the whole body of urban infrastructure, as well as the gray/blue/green public or private spaces. Given the fluid nature of contemporary life, it needs to be seen as "together with," but not in a specialized way.

The condition of "network" associated with sustainable mobility should be taken as "incremental palimpsest" by connecting human and environmental values and developing a complex system built on the large and small urban and regional infrastructure. Walking and cycling paths, public transportation systems, energy and waste management, and hydraulic plants are all common elements of the reinvention of the landscape of our cities.

All this shows the opportunity to use brownfield, in between landlocked and residual spaces. Spaces for movement, built infrastructure, and open spaces are inextricably linked.

1.1 Public infrastructure

1.1.1 Anthropic infrastructure

Infrastructure is the bedrock of economic development and urban integration in all countries. The metropolitan systems are facing a number of challenges, with regard to the provision of adequate integrated infrastructure, including:
- transportation and logistics services
- energy supply to serve increased production and to extend access to all
- access to safe drinking water, adequate sanitation and water for irrigation
- low-cost access to information and communications technologies
- meteorological services for effective and efficient planning and management of water resources, energy production, transportation services and other climate-sensitive sectors such as emergency management.

The analysis of the current situation highlights the need for a holistic and innovative approach to the planning of urban infrastructure in order to respond positively to the major challenges of a more complex future, especially when compounded by incremental competition due to increasing globalization.

The concerns about the effects of climate change on urban areas have increased the importance of accelerating investment in the green economy and sustainable development, to capture synergies between environmental protection and development of the city. Efficient infrastructure, as integrated and cost-effective prerequisites, is required to keep pace with the global economy. (Figure 1)

1.1.2 Holistic approach

Every new infrastructure, especially those linked to mobility, offers opportunities to build for a sustainable future. Bridging the identified gaps and eliminating bottlenecks are essential to making infrastructure policies pertinent to green growth strategies. Promoting sustainable approaches should optimize the performance in transportation, waste, water, and sanitation, and—to a lesser extent—the energy sector.

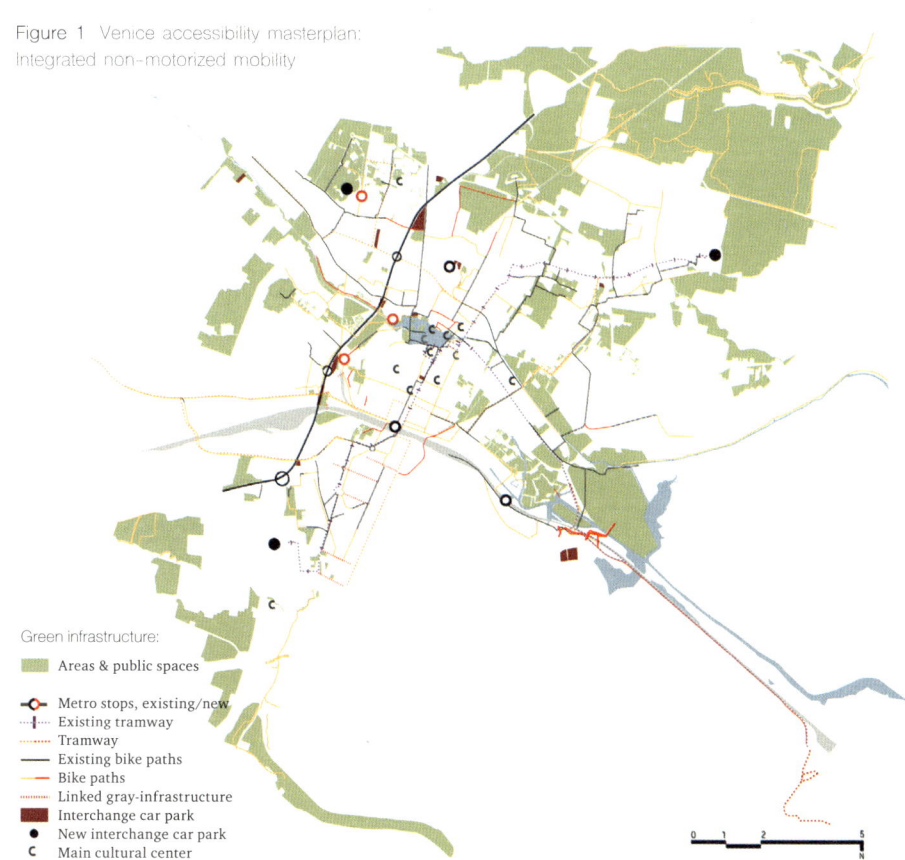

Figure 1 Venice accessibility masterplan: Integrated non-motorized mobility

Green infrastructure:
- Areas & public spaces
- Metro stops, existing/new
- Existing tramway
- Tramway
- Existing bike paths
- Bike paths
- Linked gray-infrastructure
- Interchange car park
- New interchange car park
- C Main cultural center

An ecological approach to infrastructure development is needed to unlock the potential of a city. Ecological policies address environmental, social, and economic objectives in an integrated way. An ecological approach to infrastructure planning highlights the role of climate, soil and substrate, relief, organisms, etc. The organic component of urban complexes includes the social organization of human populations. The ecological approach reviews key modes by which human populations within urban areas are differentiated, and how such differentiation affects environmentally relevant actions.

The conservation of ecosystem values and ecological functions in infrastructure development will generate wider economic, social, and environmental benefits. The protection of ecological and cultural values could not only increase the aesthetic and recreational potential of landscapes, but also have positive economic impacts on tourism-related activities. Tourists are attracted by ethical values relating to social, cultural/heritage, and environmental responsibility at the places they visit and in the products they use.

Lastly, urban ecosystems generate local services, which have a substantial impact on the quality of life in urban areas and should be addressed in land-use and climate-adaptation planning (air filtration, micro-climate regulation, noise reduction, rainwater drainage, sewage treatment, and recreational and cultural values, etc).

In this context, a holistic approach provides opportunities for incremental and transformative adjustments towards resilience/urban clime adaptation and sustainable development via effective, multi-level, urban risk governance, alignment of policies and incentives, strengthened local government and community adaptation capacity, synergies with the private sector, and appropriate financing and institutional development. Ecosystem-based adaptation is a key contributor to urban resilience.

1.1.3 Landscape infrastructure

It has become apparent that there is a need to manage urban and territorial landscape corridors as active patterns and processes in infrastructure planning. An interconnected net of natural areas, open spaces and buffers, distinguished by typologies, will be increased as a core of distinct green corridors (roads, paths, cycle tracks, parks and public gardens, riverbanks, marinas, etc) with attractive centers between the urban and non-urban areas.

With strictly integrated social, economic, and environmental revitalization in infrastructure systems, the landscape corridors form a stable environment for

Figure 2 The mobility infrastructure becomes an element of connection to the system of green areas: project of Mattuglie development in connection with the west forest of Venice

- Existing areas of environmental interest
- Project areas of environmental interest
- Cycling path
- Intervention project area
- Project areas of soft mobility

ecological communities and support effective urban policies, land-use planning strategies, and mobility networks in a sustainable urbanization.

The larger interconnected green corridors contribute to:
- creating the conditions for the preservation of biological diversity as well as the improvement of local climate and environment
- providing assets for abundant outdoor activities, and giving the inhabitants opportunities for relaxation through recreation, play, nature experiences, nature education, and cultivation
- preserving cultural and history values, and developing tourism activities.

1.1.4 Ecological infrastructure

The transition to ecological infrastructure is an essential process that will require overcoming many challenges, particularly with regard to technology, government policy, and financing. Whether the aim is to improve the energy performance of the built environment, organize carbon-free urban mobility, adapt existing networks or ensure waste collection and treatment, the entire existing infrastructure stock will have to be renovated, modified, and modernized. Greening cities therefore concerns not only the new infrastructure relating to the development of cities, but also the transformation of existing infrastructure and their maintenance.

Strategic planning of ecological infrastructure is about creating an enabling environment by delivering infrastructure requirements in order to facilitate the realization of sustainable socioeconomic development and integration. The vision of the ecological infrastructure should incorporate dual "bottom up" and "top down" approaches. This seeks to ensure that the interests of all stakeholders are taken into consideration through a local consultation and participation process, while creating a legal and regulatory environment for an effective implementation of the strategic process. (Figure 2, Figure 3)

1.2 Public spaces

There is interaction and restriction between movement and space. The public space allows all possible movements and, at the same time, influences the forms of the movements. The trails (of a person walking or cycling) may be more or less restricted and governed, but they also respond to the will of the people. A bicycle and pedestrian space is created to affect the movements of people, limiting individual human behaviors to ensure the safety of the people who use that particular space. It is a physical area established by rules and laws, giving order to the anarchy of individual behaviors. However, a public space is anarchic, which allows all movements, both predictable and unexpected. There are three main functions of open spaces strictly connected to walking and cycling issues:
- environmental and ecological functions
- social functions and human resources
- structural and symbolic functions.

Environmental and ecological functions include many so-called "ecosystem services," including:
- improving climate
- noise protection
- hydrological cycle and stormwater management
- supporting biodiversity.

Social and human functions are related to the direct use of open spaces, including:
- providing space and facilities for recreation and leisure
- promoting contact and social communication, including cultural and commercial activities
- allowing access and direct experience with nature
- producing positive effects on the health and well-being of people.

Structural and symbolic functions are related both to physical/functional open spaces and less tangible features, including:
- creating the articulation, separation and connection between different areas of "urban tissue"
- increasing the understanding of urban space
- building a sense of figuration and belonging to a specific place
- communicating identity, values, and meanings.

It is essential to remember that the widespread lack of open space in most cities means that many of these conditions are simultaneously present—although sometimes needing to be distinguished—in the same spaces. It follows a design challenge for which all the open spaces are to be carefully designed (for

1. Parking
2. Street
3. Buffer zone
4. Bike path
5. Buffer zone
6. Pedestrian path
7. Rainwater garden

Figure 3 Integrated mobility infrastructure related to natural systems: mobility as an instrument of construction of public space.
Typical section of public infrastructure development of Mattuglie

the walking and cycling infrastructure), so as to offer the largest number of uses without redundancies or clutches, keeping their potential uses open and flexible.

1.2.1 Environmental and ecological functions

The environmental and ecological value of open spaces that is achieved by the limitation of human intervention adds a "low budget" social component to the benefits for the communities and greater comfort for the non-motorized communities. Good hygrothermal balance during a hot summer period, good wind protection in winter, and an everyday noise reduction encourage the use of public spaces for both walking and cycling.
This characteristic of these spaces constitutes a first level of urban infrastructure, not heavily structured, with a high resilient potential and a quite low environmental impact. Unlike the infrastructure of a traditional type whose value is reduced through time, they have the advantage of increasing their value in the long run. The pedestrian and cyclist infrastructure includes several important components, discussed below.

● Improving climate
The open spaces favor the climatic improvement in urban areas in a lot of ways, independent from human actions:
- The open spaces dominated by vegetation contribute to maintaining a correct level of humidity in the air and soil.
- Shady areas alternating with sunny spaces favor the generation of micro-currents of air that contribute to urban comfort in hot weather.
- Ponds and fountains have a significant effect on lowering temperature.
- Open spaces located along the direction of prevailing winds can promote the circulation of air within the urban area.
- Vegetation, trees, and shrubs can be effective in eliminating pollution from particulates and providing protection from the wind.

● Noise protection
The open spaces constructed through the insertion of embankments and plant species as well as the introduction of vertical elements can be effective in the reduction of noise pollution (especially that caused by traffic).

● Hydrological cycle and stormwater management
Well-planned and implemented open spaces can bring considerable benefits to the hydrological cycle in urban areas:
- They can provide urban areas with temporary stormwater runoff storage through their form, retaining any excess fluid to avoid overloading the system of drainage and pipelines.
- They can minimize flooding, and allow the infiltration of rainwater into the soil, reducing the need for conventional drainage systems, and ensuring the right balance between evapotranspiration and thermal comfort, through adopting permeable materials instead of waterproof cover.

● Habitat for wild plants and animals
The growing importance of open spaces as ecosystems for wildlife has become particularly significant in the areas with dense human activity and decreased biodiversity caused by intensive and industrialized farming. Derelict land and brownfield sites can be considered potential habitats for animals and pioneer plants; their connection through the public space and the spaces of mobility is a real chance to return to complex biodiversity, giving an effective contribution to the promotion of environmental awareness.

1.2.2 Social functions and human resources

For people, the main conditions for using an open urban area (in addition to the mobility contents) are related to human interactions, and to the welfare and comfort derived from personal and collective experience of a space: leisure activities, games, sports, social activities, mental and physical exercises, and commercial exchange.

● Spaces and facilities for leisure and free time
The urban space can be enjoyed for games, sports, and leisure, formal or informal, active or passive. It provides:
- playgrounds for children of different ages
- the opportunity to play sports
- opportunity for recreation that does not require specific facilities.

It is important that all these facilities are available and able to attract all sections of the community.

● Contact and social communication
Open spaces are the key to the management of public life by providing important places where people of different social, cultural, and demographic groups meet and come into contact with each other. Therefore, open spaces must be considered a central part of any strategy to deepen social cohesion. There are three main factors to consider for social interaction in public spaces (Gehl, 1987):
- the growing demand from the population for open spaces
- the choice by the people to spend their free time in those spaces
- the possibility of social interaction that is reached later than these two phases.

- **Mental and physical health and well-being of people**

  The function of open spaces, in terms of psychological well-being and quality of life, has a direct impact on the health of people. In fact, health is the ability of the subjects to be in equilibrium with themselves and with their environment, and therefore to enjoy, complete physical, mental and social well-being.

1.2.3 Symbolic and relational functions

These functions are crucial for the perception of urbanity by the citizens, but in most cases are not necessarily directly related to the physical use of open spaces in question.

- **Articulation, division, and linking areas of the urban tissue**

  The role of open space in providing structure and organization in an urban area is consolidated. A city can't be separated from the surrounding landscape. Individual components of public spaces can be distinguished from each other, but they are parts of a whole. Urban open spaces are green belts and concentric rings, green corridors and slices, as well as green links of smaller dimensions, which may also take the form of simple tree-lined avenues, or roads with a low incidence of traffic. By thinking of the city as a whole with dynamic entries, other non-structural spaces are part of the system of open spaces: green infrastructure, intermediated spaces, in-between spaces, hybrid spaces that form the structure and organization of individual open spaces, and the relationship with the surrounding urban tissue. These are the spaces already naturally used by people, where we can find the available resources for increasing the non-motorized network.

- **Increasing the readability of a city**

  Being able to navigate in an urban center is crucial for several reasons, but it is more important to learn to get lost (Benjamin, 1950). Achieving a good ease of navigation and a clear sense of direction is important to ensure efficiency and a sense of well-being among the population. This deep psychological need provides a basis for the technical design and understandability of urban spaces. At the same time, the uncertainty caused by the differences provides the possibility of stimulating alternatives.

These general principles can be applied to organize and plan the elements of urban spaces—surfaces for flooring, furniture items and vegetation, lighting and water spaces—so that they can intuitively acquire meaning for the observer, who reads them with immediacy and confidence.

- **Identity and sense of belonging**

  Open spaces and urban landscape are important means of encouraging the development of a sense of belonging to a place, and the strengthening of individual and community identities.

The semantic potential of an open space is linked to the complexity of its reading and interweaving of different and existing data: like on a palimpsest, different signs—geographical and topographical, ecological, hydrological and climate, social and human, cultural and economic, symbolic and figurative—are written and erased.

Resulting from the interaction between people and place, their values are functions of how individuals, groups, or companies are able to read and interpret the stratifications through time. (Figure 4)

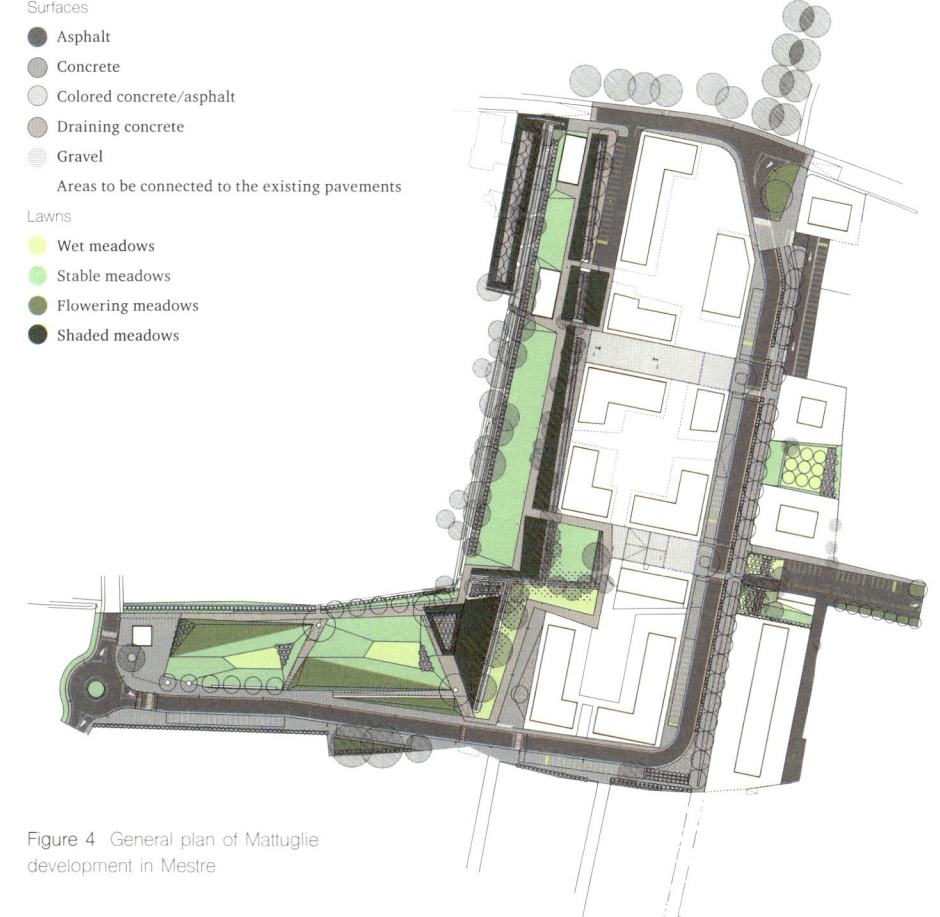

Figure 4 General plan of Mattuglie development in Mestre

2 Planning a Walking and Cycling Path System

2.1 Sustainable mobility

In 1999, the Joint Expert Group on Transport and Environment defined a sustainable transportation system as a system that:
- allows the basic access needs and development of individuals, companies, and societies to be met safely and in a manner consistent with human and ecosystem health, and promotes equity within and between generations
- is affordable, operates efficiently, offers alternative transportation mode, and supports a vibrant economy and regional development
- limits emissions and waste, consumes renewable resources at or below their rates of generation, and non-renewable resources at or below the rates of development of renewable substitutes, and minimizes the use of land and the generation of noise.

Pedestrian and bicycle transportation (also known as non-motorized transportation, active transportation and human-powered transportation) includes walking and cycling, small-wheeled transportation (skates, skateboards, scooters, and carts), and wheelchair travel. These means provide both transportation (providing physical access to goods and activities) and recreation, and belong to the system of sustainable mobility.

They represent a more individual state in which behavioral freedom, financial obligations, and an absolutely marginal collective impact on costs can coincide (according to a "low budget" policy, under which the cost of construction and maintenance of infrastructure is minimal when compared with the works required by the motorized transportation systems).

Three main items structure the concept of sustainable mobility:
- Accessibility is the value that should be maximized and equitably distributed; sustainable mobility is the vehicle to achieve accessibility.
- Mobility is clearly and increasingly an economic sector that contributes most to the greenhouse effect; at the same time, mobility is also the pressure factor that affects the quality of the local environment and public health of citizens most.
- The challenge is to create sustainable mobility at the level of individual cities and local systems, and provide the necessary contribution to the pursuit of global sustainability where possible.

2.1.1 Environmental impact

In general, the sustainable transportation attempts to reduce a city's greenhouse gases by utilizing eco-friendly urban planning, low-environmental-impact vehicles, and residential proximity to create an urban condition with greater environmental responsibility and social equity.

In order to reduce environmental impact caused by transportation in the metropolitan areas, four widely agreed points should be remembered for a low-carbon lifestyle and a healthy urban environment:
- car-free city with large pedestrian areas
- emphasis on proximity and reduced time in transit
- diversity in means of transportation, and walking-cycling urbanism
- access to public transportation by all levels of society.

2.1.2 Public transportation policies

Promotion of sustainable mobility is a complex action that must find a global, integrated, and coordinated strategy that takes a holistic approach, and addresses several contributing factors, such as:
- improving public transportation in terms of comfort, reliability, and punctuality, and preferably linking parks (better if marginal in respect to the nodes of congestion) to allow for effective change of mode of transportation
- establishing more initiatives to reduce traffic congestion: carpooling and car-sharing services, no-car days, or toll roads, park and ride, traffic lights sequences
- providing information systems that give information on the characteristics, times and services offered by the different transportation systems (urban, suburban, metro, train)
- developing controlled speed zones, limited traffic zones, and woonerfs
- developing pricing policies for parking lots to increase the operation cost of vehicles, with prices progressively growing when coming closer to the nodes of congestion.

2.1.3 Urban mobility and health

Promoting sustainable mobility policies means implementing measures in support of the demand and supply for integrated transportation systems, ensuring the movement of people and goods and, at the same time, reducing the emissions of harmful gases, noise, and energy, as well as social and economic costs. Walking and cycling, especially in urban areas, play a key role in the organization of sustainable mobility. In urban and more densely populated areas, hazards to health—generated by the mobility system—are demonstrated by:
- road accidents
- respiratory illnesses generated by fine particles
- heart disease and metabolism problems due to a sedentary lifestyle
- situations of stress and depression (especially in the elderly and children who have no autonomy of movement) due to aggressive and dangerous urban spaces.

2.2 Planning transportation strategies

2.2.1 Transportation strategy

National transportation strategies are the first level of strategy in planning walking and cycling path systems, which indicates the government's position on transportation. Every nation needs an affordable, integrated, safe, responsive, and sustainable transportation system. In a broader sense, a transportation system also aims to improve the economic, social, and environmental actions to the community through:
- improving accessibility and mobility (including walking and cycling)
- protecting and promoting public health
- developing guarantees for environmental sustainability
- increasing the safety and security staff
- promoting economic development.

The key principles include the development of:
- an integrated mix of means of transportation
- a sustainable approach in the long term
- high standards of health, safety, and security
- responses to the different needs of transportation users.

2.2.2 Integrated transportation planning

A second level is the development of integrated transportation planning that aims to embrace a range of perspectives traditionally dealt with separately, including:
- the variety of the forms of transportation (private and public, motorized and non-motorized, the different means, etc.)
- relations between mobility, and town and country planning
- the contribution of transportation policy to other economic, social, health, and environmental issues
- pedestrian and bicycle mobility as an essential part of an integrated transportation plan, and an integral part of the government vision for land transportation.

2.2.3 Other local council responsibilities

At a regional level, strategies and plans in relation to walking and cycling could include land transportation strategy (including travel demand management), walking and cycling strategies, road safety planning, growth strategy, and policy statements. Consideration must also be given to sustainable mobility, regulations and regional plans of road safety, planning documents for regional development policy, laws and programming plans, and economic and financial management.

The same strategies and plans apply at a local level, but in addition include neighborhood accessibility plans, district and city plans, long-term council community plans, asset management plans, codes of practice, design guides, and open-space access plans. Consideration must also be given to master plans and detailed traffic plans; local plans of mobility and accessibility; local plans of sustainable mobility; law and codes, regulations and design guides and "best practices"; administrative, economic, and financial programming and management.

2.2.4 Other non-transportation government strategies

The policies regarding the promotion of pedestrians' and cyclists' mobility have an important role in supporting a wide range of other activities. The actions provided for enhancing the mobility of pedestrians and cyclists should take account of and coordinate with other transportation strategies and policies in the areas of:
- health
- tourism
- cultural heritage
- environmental protection
- urban design and landscape
- planning and land development
- waste and water management
- urban regeneration
- social integration
- recreation and leisure
- economical progress
- injury prevention.

2.2.5 Local walking and cycling strategic plans

The strategic plans of local pedestrians' and cyclists' mobility (the needs of pedestrians and cyclists are different, but any combined strategies reflecting these differences should formulate joint action plans) aim to increase the number of walking and cycling trips while reducing accident rate. These two objectives are not mutually exclusive: by reducing the speed of motor vehicles and the volume of traffic, people can increase pedestrian safety and provide new conditions for pedestrian infrastructure. As a result, the strategic plans of the local bicycle and pedestrian mobility must be supported by more general strategies on traffic, road safety, and transportation.

Common elements on strategic plans can be divided into seven categories:
- **Premises**
  - mission statement and policy document
  - verification of the consistency of strategic plan with other national and local strategies
  - benefits of the strategic plan
  - the existing pedestrian and bicycle environment
  - screening of knowledge of the urban context
  - screening of knowledge and data collection of the pedestrian environment
  - analysis of critical issues and opportunities
  - study of best practices

- **Vision**—broad and strategic vision that is able to overcome issues related to only contingent questions

- **Aims**—clear statements of what the strategic plan aims to achieve

- **Actions**—description of the policies to be implemented, and actions to be taken to achieve the objectives

- **Financing**—economic and financial planning, the funding sources and the allocation procedures, the expected time schedule (including business creation and maintenance)

- **Performance monitoring and objectives**
  - a description of performance indicators to be used in monitoring the progress of the plan in achieving its objectives
  - methods and deadlines for the collection and reporting of the information needed for effective monitoring

- **Partnership/consultation**—report, consultation, and participation methodologies to achieve results with other local authorities, organizations, and communities that can support the plan.

2.3 General principles of pedestrian network planning

The main principles are based on the characteristics of a single pedestrian who is assumed to be the base unit around whom to build the space.

2.3.1 Definition of terms

A pedestrian is a person who moves on foot, or on a device equipped with wheels or skates swivel that is not a vehicle. This includes a person pushing a pram, a person on a skateboard or roller, a person in a wheelchair, and a number of other users. For simplification, a pedestrian can be placed in three main categories with similar characteristics:

- **People walking**
  - able walkers including runner/jogger, young pedestrian, adult pedestrian
  - walkers with reduced ability including elderly pedestrian, pedestrian with a guide dog, pedestrian with sensory impairments

- **On wheels**
  - person on inline skates, roller skates, skateboards, scooter
  - pedestrian with a pram

- **With impaired mobility**
  - person in wheelchair scooter, manual wheelchair, electric wheelchair
  - pedestrian with a walking frame

2.3.2 Physical space required

Pedestrians have different needs in physical spaces. The knowledge of the technological trends and needs of populations must accompany the updating of laws and regulations, especially for establishing minimum space required. For example, a new type of wheelchair may have larger dimensions, which obviously affects the design development, particularly in the regulatory aspects. So the minimum measures to consider are quite important:
- An unobstructed width of 3.28 feet (1 meter) is suitable for people with walking disabilities, but allows for only 80 percent of wheelchair-users.
- People who use wheelchairs need a clear width of 3.94 feet (1.2 meters) without obstacles.
- A width of 4.92 feet (1.5 meters) free from obstacles allows the transit of a contemporary wheelchair and a pram.

- To allow two wheelchairs to pass comfortably, unobstructed width of 5.9 feet (1.8 meters) is required.

2.3.3 Walking speed

The speed of pedestrians is influenced by:
- personal characteristics, such as age and physical condition
- psycho-motivational shift, such as purpose of walk, familiarity with the location, length of the journey, and obstacles
- physical characteristics of paths, such as width, slope, type of flooring, shelters, attractiveness, density of traveller, and possible pedestrian delays
- environmental features, such as weather conditions.

The vast majority of pedestrians walk between 2.62 feet (0.80 meters) and 5.9 feet (1.8 meters) per second. A healthy adult usually travels at an average speed of 4.92 feet (1.5 meters) per second, while older pedestrians and those with limited walking ability travel slower, at about 3.94 feet (1.2 meters) per second. It's important to consider that the current mobility scooters have speeds greater than those of most pedestrians, but can delay to maneuver between the different street levels and to work around obstacles. (Figure 5, Figure 6)

2.3.4 Abilities

Pedestrian abilities vary depending on not only the heights and age of people, but also depeneding on different physical and cognitive skills. It can also be affected by the delay or loss of capacity in terms of reaction speed, the hearing and sight loss of elderly pedestrians, as well by tiredness, fatigue or distraction. Pedestrian groups with similar characteristics can be found in some specific spatial and social conditions, such as children in the vicinity of schools, people jogging in the parks, etc.

2.3.5 Walking deterrents

The most common deterrents from walking are:
- physical environmental deficiencies
- lack of routes or their frequent interruptions
- poor quality of surfaces (deteriorated, uneven or slippery)
- obstacles on pavements (including street furniture or lighting placed incorrectly)
- lack of maintenance and hygiene (which implies the presence of waste, animal feces, foliage, or natural vegetation)
- increased distances imposed by road layouts, by physical barriers, walkways, and underpasses
- lack of a clear indication of potential destinations
- missing or unclear crossings
- poor lighting
- fast or heavy traffic
- lack of parking areas and seating
- pollution due to traffic fumes and noise
- lack of shade
- lack of shelter
- lack of interesting features along the way.

Figure 5 Local cycling mobility plan of a district in Mestre (Venice)

Figure 6 Local walking mobility plan of a district in Mestre (Venice)

© CZstudio associati and Francesco Magro

Many factors related to economic development policies, institutional and administrative choices, and criteria of social organization have helped to shape an environment unfriendly for walking.

The most critical issues are:
- spatial planning increases the distances between the origins and destinations
- higher political visibility of interventions focuses on other means of transportation
- higher priority given to other means, and less consideration of pedestrians in programs planned for other means
- inability to implement projects and programs to improve urban space by making it more inviting for walking
- difficult to quantify the social advantages brought by possible interventions focused on pedestrians
- difficult to justify the possible interventions focused on pedestrians through traditional economic criteria
- insufficient resources allocated to pedestrian programs
- lack of regulatory actions and tolerance of misconduct leading to the obstruction of passages or pedestrian areas (like parking on sidewalks).

All these factors have reciprocity and effects, but are often addressed without necessary coordination, in a distinct and individual manner, according to the fact that so many elementary procedures fail to consider the full complexity. A holistic approach is thus needed to ensure the maximum benefits for the population.

2.4 General principles of cycle network planning

The key principles for the planning of cycle networks mainly depend upon specific urban and territorial conditions that interact and affect the cyclist's choices.

These principles have great relevance in planning and design activity. The complexity of the network planning process will depend on the character of the network area, and the necessity of implementation based on the existing conditions of relevant services and structures.

The key principles include the following:
- Development of a cycling network plan has to ensure consistent routes and facilities for travel in the area, including target growth in cycling and planned changes in land use.
- A network plan can provide the basis for budget, program-scheming, and implementation priority, and can take advantage of the opportunities arising from land-use development, transportation projects, and routine road maintenance.
- The network plan also has a promotional role to encourage and facilitate cycling, particularly where the cycling network offers better accessibility and saves considerabe time compared to driving.
- The network should connect all significant destinations and attractors—schools and colleges, retail areas, primary healthcarers and hospitals, businesses, public transportation interchanges, leisure and visitor attractions, and public open space—with residential areas.
- The cycle network in most existing urban areas will predominantly utilize existing roads, through road remodification where necessary to reduce traffic volumes and speeds, and with cycle-specific measures to provide filtered permeability, improve safety and user comfort, and/or reduce delays for cyclists where necessary.
- Network implementation should prioritize routes with high travel demand, including potential demand after users switch from driving to cycling and walking.
- The communication with existing cyclists and other stakeholders is very valuable throughout network planning and implementation to identify route options and post-implementation enhancements, and to help publicize the network to potential users.

3 Types and Locations of Walking and Cycling Paths

3.1 Walking paths

A pedestrian network often passes through a variety of spaces and property:
- the road corridor, mostly public, which allows pedestrians to walk and cross streets
- public paths outside the roadway, such as coastal or riverbank walks, park paths, and pedestrian connections with other means of transportation and parking
- private land, such as the spaces pertaining to private buildings and parking lots.

The first step is to analyze the complexity and problematic nature in an in-depth way. In an integrated planning approach for the provision of new roads or changes to existing roads, it is critical to identify, understand, and act to integrate and balance the needs of all road users from the beginning.

3.1.1 Road-user hierarchy

Most roads should accommodate a wide range of users. Their conflicting needs often require a necessary balance in service level, physical space, and application limitation. Just to ensure an integrated approach, some countries use a new method to classify the roads called "road-user hierarchy" (not exclusively based on dimensional and traffic data), which strives to:
- keep all road users in mind during road planning, especially ensuring the most vulnerable users to be considered early and adequately in the process
- identify the importance of each means of transporation, and propose policies that could settle conflict
- understand the local need and select appropriate means of transportation.

In some cases, a road-user hierarchy may vary at different times of a day (such as before and after school hours, or during the day and night). This approach requires an awareness of the impacts and a vision of a wider transportation network, a clear understanding of the interaction between different means of transportation, and an analytical assessment of costs and benefits, so as to guide or prescribe various planning policies to different groups of road users.

The first phase of developing a program is to identify the importance of different road-user groups (and their positions in the hierarchy). Projects developed must be evaluated based on the costs and benefits related to users who rank highly in the road-user hierarchy. In some areas, this is the list of the hierarchy of road users from the most important to the least important:
- people on foot and people with reduced mobility on wheels
- walking people with infirmities
- cyclists/pedestrians
- public transportation users
- commercial/business users (including delivery and emergency vehicles)
- buyers who generate traffic
- visitors who generate traffic
- commuters who generate traffic.

3.1.2 Walking paths outside the road corridor

All private areas must provide a comparable, or better, pedestrian enviroment than the public road corridors.

All new buildings as well as urban regeneration are required to have a pedestrian area of high quality, which is usually better than the existing conditions.

3.1.3 Walking paths within the road corridor

Reconstruction is always preferable to a new construction, especially if, for example, there are already walking paths running through a road corridor, but they are deficient from the pedestrian point of view.

The factors that influence the possible solutions and that should be checked include:
- the reduction of traffic volume on the adjacent track
- the reduction of the speed of vehicles on the adjacent roadway
- the possibility of placing a pedestrian space within the road corridor
- the best ways of crossing road
- the possibility of walking on existing lines
- the opportunities for improving the existing space
- the interaction with the quality elements already present.

The reduction of traffic volume and speed is a condition that must be followed as a priority, as it not only results in benefits for pedestrians, but also helps to improve road safety and air quality, reduce noise pollution, and reduce infrastructure maintenance costs, while enhancing living conditions in the area.

New alignments as well as the separation of cars and pedestrians can also divert pedestrians from their usual path to accommodate and facilitate the movements of motor vehicles.

Although contrary to the scheme proposed by the road-user hierarchy, if the proposal segregation provides better access for motor vehicles at the expense of convenience for pedestrians, the order of priority may still be evaluated but not in an "ideological way," but considering both levels of convenience and real possibilities of action comprehensively. In practice, it is likely that any proposal will have to consider more than one of the conditions listed above. There may be some choices that lead to an apparently contradictory solution (for example, modal separation may discourage the reduction of the speed and volume of traffic).

3.1.4 Pedestrian environments

Variants of urban and road infrastructure that better consider the conditions of pedestrians' spaces are:
- living streets
- pedestrian precincts
- shared zones.

• Living streets

The notion of living streets recognizes that road design should take into consideration the presence and interaction of the community that inhabits it. The cars are not necessarily excluded, but the project must ensure that motor-vehicle users are aware of being in a place where pedestrians and others are more important. A living street aims to balance the needs of residents and businesses, as well as pedestrians, cyclists, and drivers, so as to improve the quality of life and accommodate a wider range of community activities. (Figure 7)

Living streets may include characteristics such as:
- traffic calming measures
- hard and soft landscaping areas
- places for social activities
- children's play areas
- seating and bar or restaurant platforms
- improved lighting
- a better interface between road and housing
- public art.

The concept of living streets can be applied to any urban road, but must be developed case by case, depending on specific factors. It is more applicable to roads that are not dominated by motorized traffic, so it tends to be adopted on urban secondary roads. There is no single solution, and it is critical to involve the community in the process of identification of problems and in the evaluation of solutions.

A design approach that considers the living streets is always desirable. The concept is particularly appropriate for all new roads, to minimize costs for a good design, as well as for existing roads that need redevelopment or restructuring works.

The functions of living streets are:
- improving security and pedestrian safety
- improving the economic vitality
- supporting community networks
- creating a sense of place and identity
- promoting cultural activities
- increasing environmental sustainability
- maintaining the ease of access to sites and services
- improving the figurative aspect of the spaces
- enhancing social interaction.

The criticizms of living streets are:
- the motorized traffic can be delayed
- the redevelopment can be expensive.

1. Sidewalk
2. Main road
3. Concrete wall
4. Pedestrians crossing
5. Vehicle entrance
6. Green area
7. Green area with gravel
8. Street
9. Parking

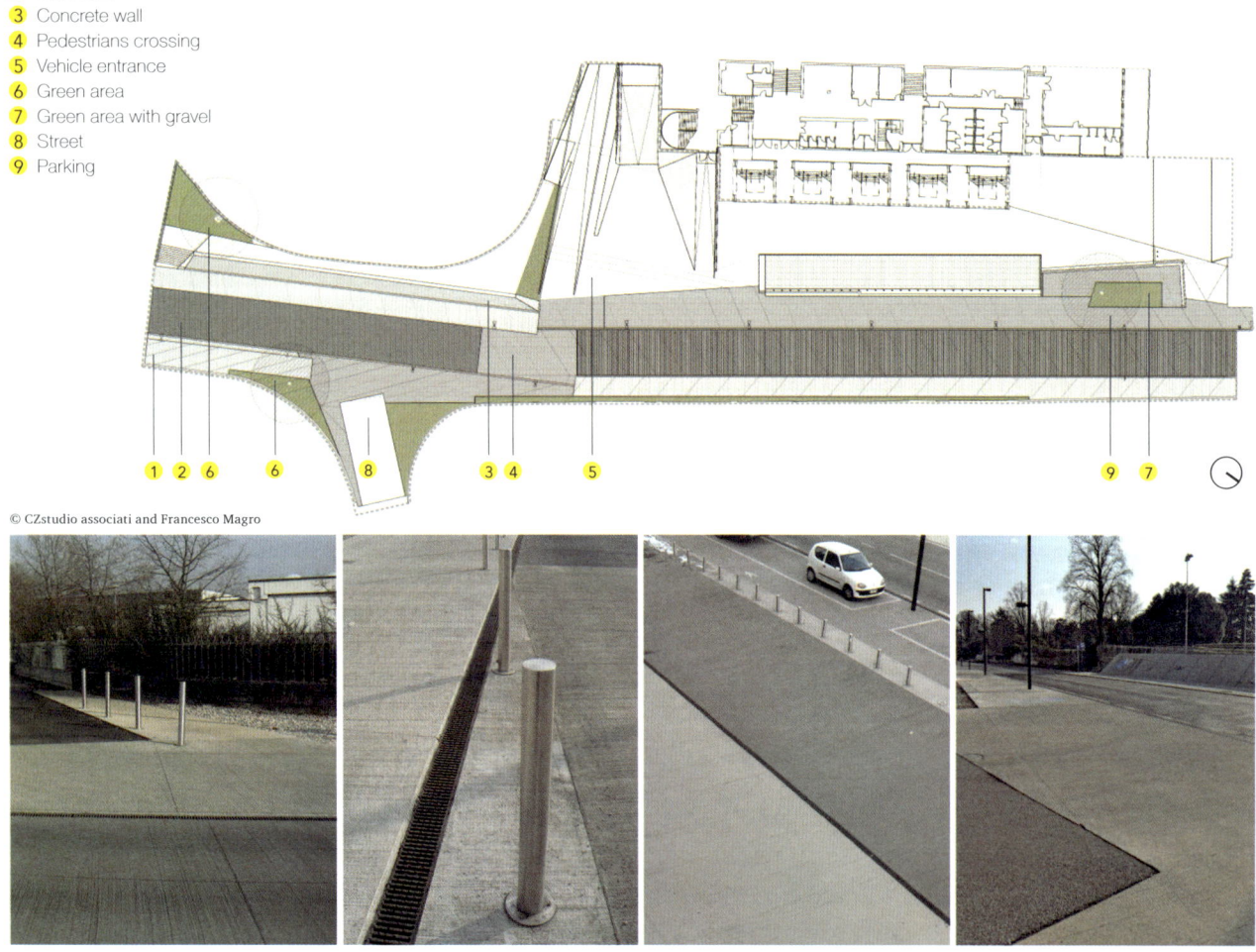

© CZstudio associati and Francesco Magro

Figure 7 Example of living street in the Pordenone Project

Flooring
- Asphalt
- Shingle
- Porphyry
- Trachite
- Mixture of lime

Mobility
- Taxi rank
- Bus stop
- P Disabled parking
- Bicycle parking

Vegetation
- Existing vegetation
- Vegetation project
- Existing hedge to cover plants

Green areas
- Public parks and gardens
- Private parks and gardens
- Possible opening to public gardens
- Periodic possible opening to private gardens

Figure 8 General plan and photos of the pedestrian path of the Eremitani Church Plaza's redevelopment in Padova

• Pedestrian precincts

Most pedestrian precincts are designed to limit traffic or close streets to traffic entirely.

There are generally four types of pedestrian precincts:
- modified street precinct: a block closed for pedestrian use only
- plaza: several blocks or urban areas closed, but the roads crossing the main proceedings remain open to traffic
- continuous paths: more isolated, and their crossroads closed
- displaced walkways: walkways developed away from usual roadside walks, making use of lanes and alleys.

The pedestrian precincts are best in urban areas where there are intense pedestrian activities due to offices and shops, high conflicts with vehicles, and possibilities of migration of motorized traffic to nearby streets.

Access is guaranteed to emergency vehicles and for loading/unloading, preferably during the early morning. Public transportation can be allowed inside reserved and protected low-speed corridors (not favorable in terms of the performances and operating costs of public transportation). Cyclists can usually be admitted as guests in pedestrian areas. Extra parking spaces may be needed to replace the spaces lost. (Figure 8)

The strengths of pedestrian precincts:
- They create the best possible conditions for free pedestrian movement and pedestrian safety.
- They bring social and aesthetic benefits; they reduce pedestrian congestion, enhance access to the retail activities, improve air quality, and eliminate noise.
- They produce economic benefits in commercial areas, helping to increase sales performance of retail business, and their competitiveness.

The criticizisms of pedestrian precincts:
- They cause trouble for traffic.
- It is difficult to convince the retailers despite their proven benefits.
- They require displacement of the routes of public transportation, which can lead to longer journey times and increase the distance to the stops.
- They lead to the closure of some routes for cyclists.
- Areas depopulate after the commercial activities end.
- They reduce the availability of parking spaces.

• **Shared zones**

A shared zone identifies, in most cases, a residential street or retail commercial area designed to give priority to residents and pedestrians, significantly reducing the dominance of motor vehicles.

Motor vehicles, including service vehicles, can access it, but must give way to pedestrians; pedestrians, on their part, should not block vehicles. Between the sidewalk and the roadway, there is no restriction.

The vehicular path is physically limited by many factors, such as trees, shrubs, and street furniture, the narrowing and changes in elevation, and parking stalls, which require slow movements with turning radius reduced. The result is a space that is difficult for motorized traffic to pass, and will only be selected when necessary. This drastically reduces the number of vehicles and ensures that drivers pay more attention.

It improves environmental conditions and road safety for the benefits of residents and clients of commercial businesses, as well as increasing open spaces to walk and play.

The shared zones are suitable for streets and areas with a low vehicular demand, and their success requires consensus and active participation of the community.

Their maximum size is determined by the necessity of ensuring access for emergency vehicles and, at the same time, the geometry of the roadway must be drawn in such a way that requires a low-speed negotiation between various users. (Figure 9)

The strengths of shared zones:
- They improve the environment through enhancing air quality, and lowering noise levels.
- They improve visual perception of the landscape.
- They reduce accident rates and severity of accidents.
- They improve social interaction and strengthen the sense of community.
- They improve individual security.

The criticizms of shared zones:
- They cause congestion of motorized traffic in the nearby streets.
- Required interventions can be expensive.
- Maintenance costs can be high.

3.2 Cycling paths

The cycle networks consist of interconnected paths and service facilities. This section describes the potential sites that can accommodate bike routes, and verifies strengths and weaknesses.

Figure 9 Shared zone of the multi-user surface of the Padua Train Station Square and Pavilions Project in Padova

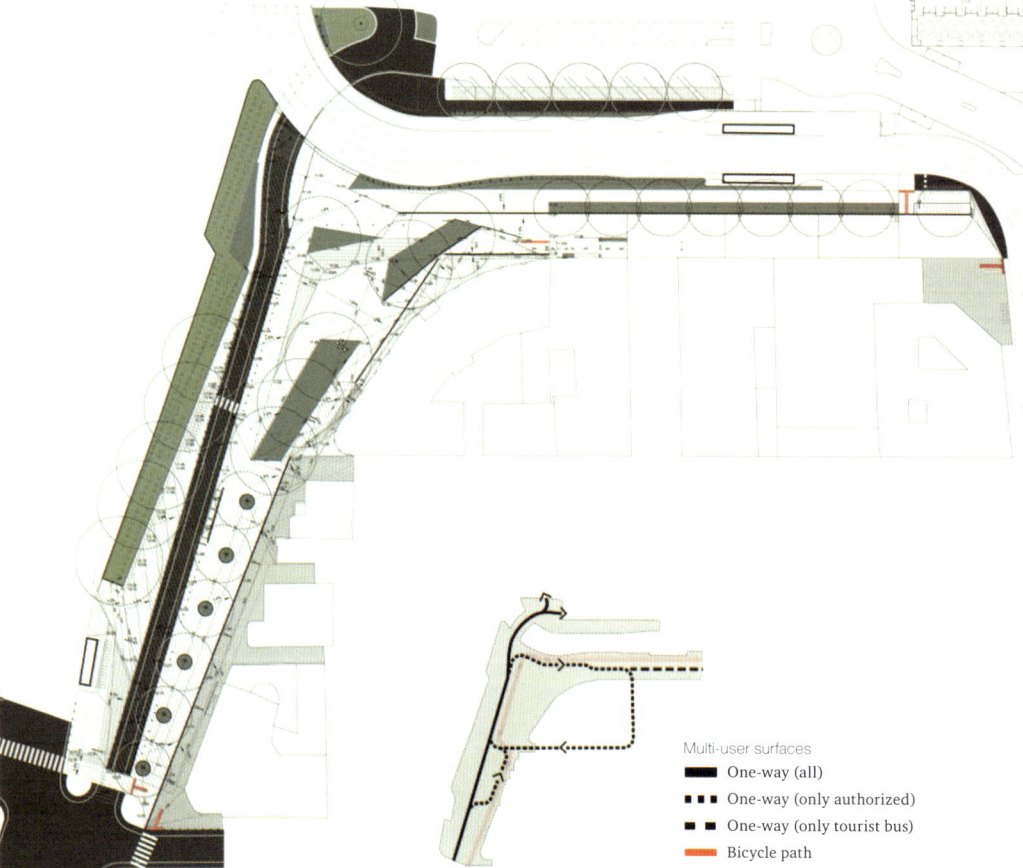

Potential sites for bike routes include:
- urban arterial roads
- urban backstreets
- urban off-road paths
- rural secondary roads
- disused railways
- watercourses
- coasts and beaches
- reserves and parks
- other locations
- public transportation networks.

3.2.1 Urban arterial roads

The urban arterial roads are the main roads of an area. Their main function is to serve transit traffic, rather than provide access to adjacent properties; along these roads, there are many important destinations.

The smaller arteries, with lower traffic volumes and speeds, generally have a single lane in each direction, and can be adapted to cyclists with basic skills, both between the intersections (known as mid-block) and at intersections.

The main streets are busier and faster, generally with more lanes. They are not suitable for beginner cyclists unless there is a separate cycling lane and more effective structures for the left turn, such as the "hook turns." To be used by most cyclists, the road should provide facilities and services suitable for cyclists with basic skills.

The strengths of urban arterial roads:
- They are often used by cyclists, and easy enough to navigate in the way of travel and turning. They need to be designed to be as safe as possible.
- Most of the arteries are often characterized by greater horizontality with respect to the surrounding local roads, as well as best surface condition and maintenance standards. They are more direct. The intersections often have signals and are constructed in a way to promote smoothness of traffic flows. They guarantee higher safety for professional cyclists because of a lower number of side roads and driveways.

The criticizms of urban arterial roads:
- The high volume and speed of traffic on these roads make them unattractive for cyclists with basic skills and for those who ride for pleasure.
- Cyclists are more exposed to automobile exhaust and air pollution.
- Even when they have bike paths, the high streets are not suitable for children and beginners.

The main constraints to the development of bicycle lanes on urban arteries consist of:
- insufficient space at intersections
- the need to maintain the standard of parking
- the potential conflicts with adjacent businesses.

3.2.2 Urban backstreets

Many cyclists who move between cities or suburban localities prefer quiet paths, especially if they do not want to blend with heavy traffic. The local roads are able to provide this condition when they extend their form in a coherent and continuous mode. Cyclist commuters use them only if they are as convenient as most direct paths, so they must be built to the same standards as those along the main road in regard to convenience and safety. Along these paths, traffic calming measures and good signage should be considered.

The strengths of urban backstreets are:
- They are the basis of the road network, especially in old cities, and are good for accommodating bike lanes.
- Backstreets are more readily available than off-road paths, and do not require extra land unless there are missing links that obstruct direct routes. Generally, all destinations are potentially connected by these pathways, which may allow cyclists to avoid particularly busy or dangerous arteries, offering a less stressful and more enjoyable path experience that is therefore also suitable for leisure or tourism.

The criticizm of urban backstreets is that to attract a significant number of cyclists, the backstreet paths need to be safer to use and cheaper to build than the main road network, which is rarely possible. Compared to the arteries, these routes usually have more intersections and more hazards from the side streets, driveways, and parked cars, which require specific protection, and signaled junctions with busier streets.

3.2.3 Urban off-road paths

Urban off-road paths are fully separated from the streets, generally within parks or inside inbuilt areas.

The strengths of urban off-road paths:
- Urban off-road paths seem to be safer due to the absence of conflict with motor vehicles, which means that they are more attractive to novice cyclists and relatively safe.
- These paths are generally placed in landscaping spaces that aim to improve the quality of the urban environment, as well as benefit pedestrian and sports lovers. They tend to encourage the movements of tourists and local cyclists who use them for recreation.

The criticizms of urban off-road paths:
- These paths are still perceived as potentially unsafe during the night and when they are poorly attended.

These paths should be well lit and well marked.
- Similar to backstreet paths, safety at intersections is a problem that requires specific protection, and signals at busy streets and other paths (even pedestrian-only paths).
- It requires a high level of design as it could be less secure than paths parallel to roads.

3.2.4 Rural secondary roads

Rural secondary roads are local roads that can provide a safer cycling alternative to rural arterials with heavy traffic or state highways (where it is better not to ride). These are country roads, sometimes paved, linking towns or villages. Rural cycling can be tortuous, requiring clear and frequent signage, especially for non-habitual users. Rural secondary roads are also good places to appreciate the landscape at low speed. These roads are valued by pedestrians and cyclists, especially when there is a speed limit for automobiles.

3.2.5 Disused railways

Disused railways are mostly found in rural areas, offering significant non-urban opportunities for touring cyclists, recreational cyclists and pedestrians at a local level. Ideally, the disused railways provide cycling of a few days at least. Connecting the main points of the public transportation, they could be justified by the presence of support structures such as shelters, toilets, rest areas and water dispensers, maintenance and support services. Because of their initial purpose, the railway networks are usually flat and direct. These disused railways can bring economic benefit to rural communities, which is less often available to communities frequented by motorized tourism, because the passing cyclists have to use the local services.

3.2.6 Watercourses

The paths close to watercourses offer a pleasant and comfortable spatial experience. They are relatively flat, and therefore suitable to be used as recreational bike paths, especially in urban areas. They also provide access to residential neighborhoods, commercial areas, and cultural facilities, and sometimes act as commuting routes. In this case, care should be taken to avoid conflicts with pedestrians.

3.2.7 Coasts and beaches

The paths along the banks of coastal towns or close to lakes and ports are heavily used for cycling tourism and for leisure, offering an experience of moving through the landscape. In many cases, climate can sometimes make them difficult to use, so some protection and service elements on the foreshore paths is recommended.

3.2.8 Reserves and parks

Natural reserves and parks are popular cycling areas for both experienced and inexperienced cyclists. These bike paths should be several miles long and offer service facilities, providing a positive cycling experience.

3.2.9 Other locations

Other possible locations include bus lanes; easements of water, gas, sewage, and electrical lines; protected areas; and abandoned streets.

3.2.10 Public transportation networks

The public transportation network extends the possibility of traveling for cyclists. Buses, trains, ferries, and planes could be considered part of the cycling network, and bus stops should be recognized in the cycle network plan.

Public transportation may be particularly useful for cyclists:
- to ensure the economic viability of longer trips
- in hilly or mountainous areas
- where there are dangerous road options, or where the roads are non-existent
- where it is forbidden to ride a bicycle (road tunnels and vehicular bridges)
- as part of the entertainment.

The success of such links also depends on:
- bicycle access at each end of the route served by public transportation
- the possibility of storing the bike while traveling on public transportation
- existence of bicycle parking at each end of the route served by public transportation (if it is not possible to transport the bike on public transportation), or if the destination is close to the interchange point with public transportation
- cost of transportation and storage.

4 Path-User Requirements

This chapter offers a comprehensive approach to maximizing the range of destinations within walking or cycling distance, improving the environment for walking and cycling, and showing individuals how these modes can effectively meet their personal needs to have the best chance of success.

People are more inclined to choose routes on foot or by bike if they perceive a pedestrian- and bicycle-friendly environment, which is convenient, safe, and pleasant, with direct connections to minimize travel time.

4.1 Pedestrians

4.1.1 Aspects of walkable communities

The "walkability" describes how the built environment is actually pedestrian-friendly. It is a useful criterion for evaluating the characteristics of an area or a route, although it results from a totally subjective interpretation of individual conditions. Some key features may be identified to provide an initial description of a walkable community.

- **Connected**
The network must ensure direct access for pedestrians to the destinations they want to reach. The paths must be well connected to public transport and the surrounding networks.

- **Clear**
The network must be clearly identified, published in local maps, and described through applications for smartphones, so that the first-time visitor can navigate comfortably.

- **Comfortable**
The paths must not be contaminated by excessive noise or traffic fumes. They must be quite large, with smooth surfaces and gentle slopes. Preferably, they should provide shelters and resting places.

- **Convenient**
The paths must be continuous, in good condition, without obstacles, for example, the users are not delayed by other users or by road traffic.

- **Pleasant**
The pedestrian spaces should be pleasant, interesting, not chaotic, and very clean, with a quality that encourages continued use and social interaction.

- **Safe**
The road crossings and those at the driveways should be safe from traffic. The materials and surface treatments must be designed to avoid slipping and falling.

- **Secure**
The pedestrian environment should discourage anti-social and criminal behavior through a design that uses the principles of crime prevention.

- **Universal**
The routes must be equipped with devices to ensure access for people with limited mobility through gentle slopes, visual contrasts, sounds, and specific tactile characteristics.

- **Accessible**
The most common destinations have to be at a contained and walkable distance.

4.1.2 The importance of urban form

Urban form refers to how the settlements are designed and structured, which kind of development is allowed, and where and how different areas are connected.

The shape of a place affects the reasons for a trip and the attractiveness for walking. There are three elements that will affect urban form, which can be evaluated by their relationship with pedestrians:
- pedestrian permeability
- connectivity
- strategic planning.

- **Pedestrian permeability**
It's a measure describing how an environment is accessible to pedestrians, free from obstacles, and viable because of frequent and comfortable interconnections. A permeable form allows for:
- pedestrian priority at traffic lights and pedestrian crossings
- pedestrian priority at the entry with the side streets
- possibility of using closed-off routes
- traffic-calming measures, low-speed zones, and pedestrian common areas.

- **Connectivity**
It's a measure describing how the pedestrian network is integrated with the origins and destinations of

possible shifts (including the network of public transportation):
- The pedestrian network connects the more important and often-used terminals, such as schools, shops, supermarkets, parks, public spaces, and service centers.
- Special attention is addressed to the interface between urban attractions and pedestrian network, such as the provision of shelters, shaded seating, and pedestrians' signage.
- The environment closer to public transportation and interchange nodes is intensively developed and more pedestrian-friendly.

- **Strategic planning**
It's a measure describing how policies and local strategies are able to encourage pedestrian traffic as a means of transportation by:
- coordinating between land-use planning and transportation planning
- ensuring that local development policies promote walkability
- ensuring the urban development policies encourage the increase of population density around transportation nodes
- implementing a program of regular monitoring of foot traffic
- coordinating and integrating the management of open spaces, parks, and roads, both in terms of route planning, lighting, and signage.

Since walking is a slower mode of moving, any deviation from the most direct route will cause more inconvenience than those that might be caused to other modes of transport. Pedestrians benefit more than others from the permeability assured by smaller networks.

The appropriate distance of the terminals of a pedestrian network could be judged by comparing the average distance covered in an hour by different modes of transportation: walking 31.07 miles (5.4 kilometers) per hour; cycling 12.43 miles (20 kilometers) per hour; driving 33.56 miles (54 kilometers) per hour.

Walking is facilitated by compact urban models, which have an interconnected network of roads, with high-density populations, and a pattern of sociability that favors the search of work, as well as a way of living that prefers the housing near shopping and neighborhood services, close to the transportation hubs and, if possible, to the open spaces and parks. Walking is a good travel option in these areas because of the short distance between the origins and destinations. It becomes safer and easier for people to walk to the places of work, shopping, education, leisure, and services.

The road network should not necessarily be as permeable as the pedestrian network. The highly connected road networks are critical for safety at intersections, which therefore require traffic-calming measures.

Careful urban planning can change the behavior of motorists, cyclists, and pedestrians, and reduce the dominance of motorized traffic. Even the creation of a multi-purpose space in the new developments can prevent a road network from using uncoordinated and unrelated traffic-calming devices.

In all projects of open spaces, a higher priority must be given to pedestrians rather than cars. The quality of an urban scene is obviously more relevant to walking experiences while focusing on both detailed views and general aspects. For this reason, when the flow of vehicles is small, it is generally acceptable to reduce the presence of traffic engineering devices and focus on urban design through the possibilities given by the architectural and landscape tools.

The vision of a road is both functional and responsive to traffic engineering. Roads and public spaces should be pleasant, engaging, and stimulating. It is important to foster an environment where impediments to pedestrians are introduced only when absolutely necessary.

4.1.3 Policies affecting walkable communities

Urban and regional planning and pedestrian infrastructure have a significant impact on walkability.

- **Land-use planning (in existing areas)**
A wide range of places of interest is available at walking distance in historic centers, local communities, and suburbs.
- The destinations of the city center are equipped with a wide variety of shops, offices, services, entertainment, and public spaces.
- In local communities, the destinations include a range of businesses, community and civic services, schools, and medical services.
- All homes are within walking distance from a public space and frequent public transportation.

The walking distance here must be set at the local level and based on travel times in relation to the urban scale.

- **Land-use planning (for new developments)**
- The policies and local plans must specify the permeable network, and not allow layout offering exclusively winding paths and dead ends that have no alternative outlet for pedestrians.
- A range of services and relevant destinations are designed close to the origins to promote the development of functional mix, especially near roads that could also be accessed by public transportation.

- Crime prevention through environmental design principles should be applied to all new developments.
- Attention must be paid to the expectation and location of parking.

- **Pedestrian infrastructure**
  - Pedestrians have to be considered from the earliest planning stage of infrastructure. Adequate pedestrian services are established through the road-user hierarchy and provided by physical network paths.
  - The entire networks provide equipment for people with limited mobility and cognitive abilities.
  - The speed of motor traffic is handled by pedestrian-oriented design and regulations. In areas of great importance or with a high pedestrian density, or in neighboring areas, the traffic speed is determined by pedestrians, or sidewalks are specifically designed as an alternative to offer a higher level of security and convenience.
  - Pedestrians are generally placed at the top of the road-user hierarchy, because their needs must be met, ensuring a high level of security and accessibility.

4.1.4 Personal security issues

Problems related to the individual security can constitute a major obstacle to the creation of a "pedestrian community," especially during the night and in rundown places. When perceiving dangers, people may change their travel behavior, and avoid going out at night, staying away from routes, which inhibits independent movement. Security issues can also create a barrier to access to public transportation on foot.

The main issues related to personal safety should be considered under three general conditions:
- The site must be readable, with pedestrians able to understand and see their immediate surroundings and the people in front.
- The pedestrians should be visible to others, especially other pedestrians.
- Any route should have alternative routes to avoid potentially dangerous situations.

It is particularly important to:
- provide adequate lighting
- avoid design solutions that can create hidden areas
- give pedestrians the maximum number of route choices
- maximize "natural surveillance"—the number of people entering an area of houses, streets or places of amusement
- provide proper signage
- provide a clean and "cured" space.

Personal security issues can be better addressed if we consider:
- strategic plans for walks
- accessibility of local plans
- pedestrian walkability surveys, with audits
- safe routes to school
- walks to the school bus
- pedestrian lighting services
- planting and maintenance of vegetation
- maintenance of paths and adjacent areas
- promotional campaigns for walking.

4.2 Cyclists

4.2.1 Aspects of cycling communities

The "cycling community" refers to the extent to which the built environment is bicycle-friendly. It's a useful way to evaluate the characteristics of a zone or a path, although it is clearly established through personal criteria.

The following main features may provide a description of a possible bicycle-friendly community.

- **Security**
  The bike paths are safe, ensuring personal safety, and limiting the conflict between cyclists and other path users. Safe cycling is influenced by the speed and volume of motorized traffic, so it may be convenient to separate bike paths from roads. Cyclists should always have a dimensioned space to ensure a high level of personal security. At the same time, it is crucial to ensure safety at the points of conflict, particularly at intersections, through careful design. Even careful design of street lighting and signage can help to improve personal safety.

- **Comfort**
  The bike paths should be smooth (with roughness to ensure a non-slip surface), clean, and free of obstacles, with a reduced slope, and a design to make travel easier. In the design phase, provisions should be considered to reduce the impact of rain, wind, and heat, for example:
  - walls, embankments or hedges along the tracks (without diminishing the visibility and safety criteria)
  - shaded paths
  - construction that can be easily repaired at the critical sections.

- **Immediacy**
  Generally, bike paths should be direct, with parking spaces guaranteed in convenient locations, to attract the users by helping them to save time and energy. Excessively indirect cycling routes, delays, or lack of parking can lead users to choose a more direct road with greater risks, or simply abandon cycling.

- **Consistency**
The bike paths should be continuous and easily recognizable, connecting all the most interesting origins and destinations, and offering a uniform and consistent standard (also in terms of traveling speed).

- **Attractiveness**
The bike paths are built to interact with the environment, and enhance the attractiveness of the landscape. In urban areas, they should be always integrated into the design of public spaces. The path should be well designed to fit into the urban landscape, and should also provide information (info point, QR code, wifi), which contributes to a pleasant cycling experience.

- **Adaptability**
Special attention should be paid to the adaptability of infrastructure where a substantial increase of cyclists may be expected.

4.2.2 Cycling trip types and requirements

For planning purposes, the types of cycling journeys can be grouped into:
- neighborhood cycling
- commuter cycling
- adult sport cycling
- recreational cycling
- touring cycling.

- **Neighborhood cycling**
Most neighborhood cycling refers to trips to local schools and shops and children playing on their bikes. Services in support of cycling should therefore pay special attention to the needs of beginners. In residential neighborhoods, the speed of cycling is generally below 9.32 miles (15 kilometers) per hour. However, the cyclist may have to cross arterial streets and short sections of the primary cycle network to reach local destinations. The highest priority is given to ensure a safe environment for children and beginners on local roads and around retail stores and schools.

These cyclists prefer:
- the highest degree of safety
- comfort and personal safety
- low speeds and reduced traffic volumes
- good separation from traffic when the destinations are on busy streets
- a minimum slope
- safe crossings on busy roads
- road signs
- a secure parking lot next to the destination
- good lighting
- protection from the wind and rain
- the bike path integrated with the surrounding landscape.

- **Commuter cycling**
Most cycling commuters are high school students or adults. Most of their trips are developed on the arteries or other major cycling routes.

The regular commuters move generally at a speed of 12.43 to 18.64 miles (20 to 30 kilometers) per hour, while the average length of the journey for cycling commuters is about 3.11 miles (5 kilometers). Most of them prefer to choose a faster route rather than a safer, more comfortable and attractive one. They are the main users of the primary cycle network.

It's important to note that the projects designed in accordance with the needs of experienced commuters are usually less attractive to new users with less confidence; therefore, the services in support of cycling should not only meet the needs of expert cyclists, but also keep in mind other commuters.

These cyclists prefer:
- high-quality road surfaces
- direct and consistent routes
- minimal delays
- infrastructure dedicated to them
- intersections that minimize conflicts with other commuters
- good lighting
- guarded parking lots near the destinations
- facilities such as lockers and showers.

- **Adult sport cycling**
Serious cyclists often travel faster than 18.64 miles (30 kilometers) per hour. They are experienced cyclists ready to claim their road space and less available to negotiate it. They usually travel long distances, especially along high streets or rural roads, and can explore challenging terrain. Often they are traveling in groups of two or more, and love to travel side by side.

These cyclists prefer:
- high-quality road surfaces
- minimal delays
- wide roads
- physically challenging paths and demanding gradients.

- **Recreational cycling**
Recreational cycling is usually less constrained by time. The cyclists usually have very different skills and experience. The most popular cycling destinations include routes along the rivers, coasts, and reserves, as well as attractive locations with a low volume of traffic and low speeds. These cyclists prefer:
- comfort
- good surfaces
- minimum slope
- a high degree of safety and security

- pleasant, attractive, and interesting routes
- protection from the wind and rain
- a parking lot where there are facilities or attractions to visit during the trip.

• **Touring cycling**

Touring cyclists travel long distances, and often carry camping equipment and provisions. Generally, they are experienced, and travel in pairs or in groups.

These cyclists prefer:
- routes to pleasant, attractive, and interesting places
- adequate open side along the way
- high-quality road surfaces, although some may try to travel on backstreets with little traffic
- facilities such as rest areas, water dispensers, toilets, and shelters.

5 Design Criteria for Walking Paths

The main design criteria for walking paths:
- Public and private spaces and roads have to be equipped with sidewalks where there is pedestrian need.
- The dimensions and geometries of sidewalks should ensure accessibility for all.
- Construction materials should take into consideration safety, beauty, and affordability.
- Spatial design should be considered to integrate signage, street furniture, and lighting.
- The driveway should be identified prior to the design of urban space to achieve a more integrated plan.
- Solutions to potential conflicts between different modalities in shared spaces should be provided through the design of the open space.
- The project should be strictly linked with the public transportation stops.

5.1 Walking path zones

Most of the pathways are formed between the edge of the driveway and the front of the buildings or the limits of the adjacent private areas. There are four distinct areas, and it is important to distinguish between their overall width and the width of the area for pedestrians (the direct path).

The elevated routes are insurmountable since they have a height exceeding 2.95 inches (75 millimeters) from the road. The maximum height of a sidewalk cannot exceed 5.91 inches (150 millimeters).

5.1.1 Zones of a walking path

Thoroughfares, where pedestrians normally choose to walk, should always be kept clear of obstructions. Pedestrians tend not to use frontage areas, as they may contain retaining walls, fences, shop-related elements, and vegetation. They can also be busy with people coming out of the building.

Curbs are often made with precast concrete elements (5.91 inches [150 millimeters]) or steel plates (0.39 inches [10 millimeters]) to ensure they fulfill their many functions, including:
- defining the limit of pedestrian space
- preventing the flow of water from the road onto the sidewalk
- preventing vehicles from using the sidewalk
- serving as a tactile element for pedestrians with visual impairments.

Street furniture areas serve many purposes, including:
- providing space to install signals, lighting columns, bollards, billboards, and parking meters
- displaying plants
- creating a psychological buffer between motor vehicles and pedestrians
- protecting pedestrians from water spray of moving vehicles
- providing space for the ramps connecting the sidewalk and street at driveways.

5.1.2 Width of zones

The width of walking paths depends on the urban zones where they appear, the flow of people, and the importance of connected elements.

In general, the areas that require more extensive pedestrian areas are those with:
- a proximity to high-speed vehicles, and/or high volumes of vehicles
- extensive crossing areas
- high pedestrian volumes, and/or high number of pedestrians who are stationed on the sidewalk.

● Table 1 Width of zones on a typical footpath

Location	Maximum pedestrian flow	Zone				Total
		Curb	Street furniture①	Through route	Frontage	
Arterial roads in pedestrian districts	80 p/min	0.03-0.05 ft (0.01-0.15 m)	3.94 ft (1.2 m)	≥7.87 ft (2.4 m)	2.46 ft (0.75 m)	14.3-14.76 ft (4.36-4.5 m)
CBD						
Alongside parks, schools and other major pedestrian generators						
Local roads in pedestrian districts	60 p/min	0.03-0.05 ft (0.01-0.15 m)	3.94 ft (1.2 m)	5.91 ft (1.8 m)	1.48 ft (0.45 m)	11.35-11.81 ft (3.46-3.6 m)
Commercial/ industrial areas outside the CBD						
Collector roads	60 p/min	0.03-0.05 ft (0.01-0.15 m)	2.95 ft (0.9 m)	5.91 ft (1.8 m)	0.49 ft (0.15 m)	9.38-9.84 ft (2.86-3.0 m)
Local roads in residential areas	50 p/min	0.03-0.05 ft (0.01-0.15 m)	2.95 ft (0.9 m)	4.92 ft (1.5 m)	0.49 ft (0.15 m)	8.4-8.89 ft (2.56-2.7 m)
Absolute minimum②		0.03-0.05 ft (0.01-0.15 m)	0.0 ft (0.0 m)	4.92 ft (1.5 m)	0.0 ft (0.0 m)	4.95-5.41 ft (1.51-1.65 m)

① Consider increasing this distance where vehicle speeds are higher than 50 km/h (31.07 miles/h).
② Only acceptable in existing constrained conditions and where it is not possible to reallocate road space.

Based on Table 14.3 in New Zealand's *Pedestrian Planning and Design Guide*; amended with Italian-EC law's standards

● Table 2 Installing passing places

Reason	Passing place dimensions	Location and spacing
Wheelchair users	minimum width 5.91 ft (1.8 m) minimum length 6.56 ft (2 m)	at least every 164.04 ft (50 m), where the sidewalk is less than 4.92 ft (1.5 m) wide
Passing pedestrians	minimum width 5.91 ft (1.8 m) minimum length 16.4 ft (5 m)	as required, where pedestrians may wait

Based on Table 14.4 in New Zealand's *Pedestrian Planning and Design Guide*; amended with Italian-EC law's standards

If the flow of pedestrians per minute (p/min) exceeds the maximum (shown in Table 1), then a wide path needs to be considered.

5.2 Passing places

Passing places should be provided where the clear width of the path is constrained to less than 4.92 feet (1.5 meters) wide. They consist of enlargements point, and should be made where it is not possible to widen the path for a relatively long distance, but they should never be used as a low-cost alternative to a wider path (Table 2). The advantages of passing places are:
- two wheelchairs can pass at the same time
- moving pedestrians can pass by any people who are stopping to use a crossing or waiting for public transportation.

5.3 Overhead clearances

To prevent head injuries, the sidewalks must be free from overhead obstacles. They must be free from any type of object (street furniture, vegetation, signs, and elements attached to buildings) from the ground to a certain height, and across their full width. The ideal obstruction-free height is 7.87 feet (2.4 meters); the absolute minimum is 6.89 feet (2.1 meters).

5.4 Cross fall

Cross fall is the slope of a sidewalk perpendicular to the direction of travel. Paths need a certain slope for drainage of the water, which must always be between 1 percent and 2 percent at most, and is generally towards the street.

5.5 Surfaces

All surfaces on which pedestrians walk must be stable and non-slip, even when the surface is wet. Sudden changes in height on flat surfaces must be less than 0.2 inches (5 millimeters) to minimize the risk of tripping, while undulations on surfaces must be less than 0.47 inches (12 millimeters). (Figure 10)

Both the above points are followed where the maximum deviation of the surface over 19.69 feet (6 meters) of straight edge is less than 0.2 inches (5 millimeters). This also prevents puddles forming. Curved channels for drainage should not be placed within the path. Brief, sudden changes in the surface, such as individual steps, should be avoided as they are unexpected and my cause pedestrians to stumble.

5.5.1 Surface advantages and disadvantages

Different types of path surfaces have advantages and disadvantages, as listed in Table 3.

Figure 10 Different types of surfaces

Table 3 Surface advantages and disadvantages

Surface	Advantages	Disadvantages	Design issues
Concrete and asphalt	require minimum ongoing maintenance; any maintenance is inexpensive; surface can easily be reinstated if removed; provide longest service life	can be aesthetically unappealing; asphalt can be confusing for pedestrians as it is associated with road surfaces; asphalt can sink and produce protrusions, especially at curbs	texture with a broom finish (perpendicular to the direction of travel) to enhance friction and improve drainage; concrete can contain colored oxides; joints between units shall be less than 0.59 inches (15 mm)
Stone pavers and unglazed brick	highly decorative; easy to replace if damaged; easy to reset if displaced.	small units can move independently and create a tripping hazard; can be difficult to maintain cross falls; can cause vibration to users; some pavers or joints are susceptible to moss	consider stamped or stained concrete instead, and joints between units shall be less than 0.59 inches (15 mm); needs a firm base (preferably concrete slab); ensure good installation and regular maintenance to prevent moss growth and minimize/reset displaced pavers
Split-face stone, cobblestones	highly decorative	not easily crossed by the mobility-impaired or pedestrians wearing some fashion shoes; prone to moss and weed growth	avoid use in thoroughfares; can be used to delineate places to walk, and other areas of the sidewalk
Loose surfacing, such as exposed aggregate, gravel and bark	inexpensive to install; can be aesthetically pleasing; can fit well in rural environments	can cause severe problems for the mobility-impaired if not well compacted; requires significant maintenance; very prone to weeds	avoid use in thoroughfares unless there is an extremely high aesthetic justification; use to manage vegetation and street trees only (and take measures to prevent materials spilling into the thoroughfare)
Tactile paving	provide a positive tactile wayfinding cue for the vision-impaired.	can be aesthetically unappealing	should be used only in specified locations

Based on Table 14.8 in New Zealand's *Pedestrian Planning and Design Guide*; amended with Italian-EC law's standards

5.5.2 Materials: performances

Concrete and asphalt surfaces are generally considered best for sidewalks, although paving stones or bricks can also be used. Wood requires a lot of maintenance, but can be used for very large surfaces. Particular attention must be paid to the selection of pavement materials, which should always be able to withstand solid use by displaying some or all of the following characteristics:
- bending strength under load after freeze/thaw cycles
- compressive strength after freeze/thaw cycles
- slip resistance
- water absorption at atmospheric pressure
- abrasion resistance.

The combination of different materials must be adopted with the aim of obtaining the best outcomes regarding safety, cost of construction and maintenance, and environmental comfort. Desired outcomes include:
- ease of manufacture, cleaning, and maintenance
- drainage capacity
- reduced heat absorption.

5.5.3 Materials: urban comfort

The increasing population in cities has great impact on the distribution of energy, environmental conditions, and public health. One consequence, amplified by climate change, is the strengthening of the urban heat island (UHI), which is represented by the rising of temperatures in cities compared to rural areas. The heat island is caused by human activities, which have replaced vegetation and permeable surfaces with built surfaces characterized by high levels of solar absorption, high impermeability, and thermal properties that increase the amount of stored thermal energy. Many other human activities also contribute to the heat island effect: public and private transportation; heating systems in buildings; the hot air released from the cooling systems, etc. Several studies have quantified this phenomenon, and the intensity of the heat island has been recorded at over 53 °F (12 °C).

The UHI has a severe impact on the performance and energy costs of buildings, as well as carbon-dioxide emissions. It also triggers the intensification of health problems with smog and pollutants during heat waves.

The peak temperature on the surface of building materials can reach up to 86 °F (30 °C), higher than that of air, and this is a great problem, especially for walking and cycling communities. The thermal energy stored is then released into the environment, resulting in a temperature increase. Streets, sidewalks, parking lots, pedestrian areas, and bicycle paths play an important role in this energy balance, as they may represent between 30 percent and 45 percent of the urban footprint. The values of solar reflectance for materials representing urban structures are typically 5 percent for newly laid asphalt (asphalt covers the surface uniformly); 15 percent for older asphalt; and 25 percent for concrete.

Cool materials are a special category of materials and components characterized by high solar reflectance that induces a minor rising of surface temperature when they are subject to solar loads. This technology aroused great interest for the mitigation of urban temperatures because it has very high thermal emissivity. This feature allows the thermal energy stored during the day to be radiated into the sky at night, contributing to the decrease in temperature of the entire urban area.

Cool pavements are designed for urban areas, which requires three factors to be taken into consideration:
- Urban conditions, or the massive use in construction of a road surface used both by pedestrians/cyclists and motorized vehicles, imply a larger thermo-physical study on wear phenomena, convection, variability of shading, and shape of a space. All this makes modeling them difficult.

- The surface temperatures of asphalt and concrete used for paving are affected by both radiative and thermal characteristics.
- The road surface must fulfill different functions within an urban area. Their uses range from walking and cycling paths to busy motorways.

The cool pavements, although originally designed to control stormwater and characterized by a water-permeable structure, are emerging as potential cool materials. These surfaces allow for storing air, water, and water vapor in their interstices. They are classified as permeable surfaces: porous asphalt, draining concrete, and some special flooring for sidewalks. Due to the permeability and structural requirements of different types of traffic, these floorings must be supported by both an efficient design and a proper installation.

Under particular climatic conditions, these materials should be able to limit the rise in surface temperature caused by solar loads, through evaporation. In fact, the water stored during the rainfalls inside the cavities and porosities of their structures serve as reservoirs of latent heat by vaporizing the outer layers of the materials during irradiation. This phenomenon is quite similar to evapotranspiration operated by vegetation.

The surface conformation of these products can also affect the ambient temperature. A high number of cavities will result in more shadow zones distributed on a wider surface. These conditions may limit the processes of heat transfer to the substrate by subjecting the outer surface of the heating due to solar radiation, but reducing the accumulation of heat and the consequent release during the night hours. A larger surface area also facilitates a more effective mixing, encouraging the processes of convective heat transfer.

Figure 11 Different types of grates and covers

Figure 12 Permanent planting of a road

5.6 Grates and covers

When possible, the covers of manholes and grilles must be located within the area of street furniture. If this is not possible, they can be placed at the edge of the thoroughfare, and must be flush with the surrounding surface and be non-slip, even in wet conditions.

To minimize the risks for pedestrians, it is recommended that the slots of grids have a width less than 0.59 inches (15 millimeters), and are positioned perpendicular to pedestrians' main direction of travel. (Figure 11)

5.7 Landscaping

Landscaping, which is highly varied across different climatic zones, is to create a visually attractive space

Adolf B. Horn Avenue—AGRAZ Arquitectos

and a buffer between the sidewalk and the road. The presence of plants can give drivers the sensation of being in a "restricted space," inviting them to reduce speed, while providing shade and shelter for pedestrians from wind and sun.

Permanent planting should be located within the urban area, consisting of trees, flowers, shrubs and groundcover, grass, and herbaceous species. The species must be selected with care to ensure that they fit the urban scene and can easily adapt to the environment (Figure 12). The following characteristics are particularly important:
- The root systems do not damage the buried networks or raise the pedestrian areas.
- Vegetated pergolas do not interfere with overhead lighting.
- Vegetable elements must not obstruct the view of pedestrians and motorists at any time of the year.

Trees with branches lower than 6.56 feet (2 meters) and shrubs taller than 1.64 feet (50 centimeters) should be avoided.
- The plants must be able to survive with minimal maintenance (in the drier areas), and preferably without the need for irrigation after the start-up period (at least two years).
- The vegetable elements should not be arranged in such a way as to constitute shelter for criminal or anti-social behavior.

It's good to check the landscape project is consistent with policies relating to road safety, for instance, the shapes of the designed elements and their locations should not create danger for vehicles on the road. Outside of moderate traffic areas, if the road is bordered by insurmountable curbs, it is prudent to use collapsible or breakable furniture and landscape components if they are up to 13.12 feet (4 meters) from the edge of the nearest traffic lane.

5.8 Ramps and Steps

A pedestrian path should be treated as a ramp if the mean gradient is greater than 8 percent. Stairways and ramps must be straight if possible (Table 4, Table 5). Curves in ramps and stairs are not recommended because:
- they become more difficult for people with limited mobility to use
- for ramps, the gradients between the internal and external edges are different

● Table 4 Design features specific to ramps

Parameter	Range/value
Surface	should comply with the same best practice as other pedestrian path surfaces
Width	3.94 ft (1.2 m) absolute minimum, preferably 5.9 ft (1.8 m) between handrails); if more than 6.56 ft (2 m), a central handrail should be provided
Maximum length	preferably less than 164.04 ft (50 m); absolute maximum 426.51 ft (130 m).
Maximum height	absolute maximum 11.48 ft (3.5 m)
Maximum cross fall	2 percent (but no cross fall normally required)
Mean gradient	no greater than 8 percent
Maximum gradient	generally no greater than 8 percent; in highly constrained conditions, greater gradients are tolerated but only over short distances: • a gradient of 10 percent is permitted over a length of 4.92 ft (1.5 m) • a gradient of 15 percent is permitted over a length of 1.64 ft (0.5 m)

Based on Table 14.11 in New Zealand's *Pedestrian Planning and Design Guide*, amended with Italian-EC law's standards

● Table 5 Design features common to both ramps and steps

Feature	Purpose	Location	Design issues
Landing	accommodate changes of direction after the ascent/descent is completed; ensure that pedestrians emerging from the ramp/steps are clearly visible	top and bottom of every ramp or flight of steps	at least 3.94 ft (1.2 m) long, 5.9 ft (1.8 m) preferred; extend over the full width of the ramp/steps; keep clear of all obstructions; gradient should be 0 percent, always less than 2 percent
High contrast or different roughness material	enable people to detect the top and bottom of the ramp/steps	edge of the landings, adjacent to the ramp/steps; on the edge of each step	on the edge of each step; should cover the full width of the steps/ramp; on steps, it should be 2.17 inches (55 mm) deep
Tactile paving	help vision-impaired people to detect the top and bottom of the steps or steep ramps	edge of the landings, adjacent to the ramp/steps	install tactile ground surface indicators
Signing	inform pedestrians of the impending change in levels; provide directions to an alternative route where available	top and bottom of every ramp or flight of steps	no additional requirements to normal pedestrian signage
Handrails	provide a means of support, balance and guidance; provide a means of propulsion for some pedestrians	continuous over the whole route and provided on both sides	handrails should be from 1.2 to 1.97 inches (30 to 50 mm) in diameter and sited at least 1.97 inches (50 mm) from any surface; should extend by at least 1.2 inches (30 mm) into the top and bottom landings, and return to the ground or a wall, or be turned down by 3.93 inches (10 cm); sited 3.28 to 3.6 ft (1 to 1.1 m) above the step pitch line or ramp surface; secondary handrails may be considered at a height from 2.46 to 2.79 feet (75 to 85 cm).
Rest areas	allow pedestrians to recover from their exertions; make changing direction much easier	frequency depends on the height gained (or lost); a rest area is required every 2.46 ft (75 cm) change in height for the ramp to remain accessible to wheelchair users (see detail in Table 6).	at least 3.94 ft (1.2 m) long, 4.92 ft (1.5 m) preferred; covers the full width of the ramp/steps; gradient should be less than 2 percent

Based on Table 14.10 in New Zealand's *Pedestrian Planning and Design Guide*, amended with Italian-EC law's standards

Table 6 Rest areas and ramps

Gradient	>5%	5%	6%	7%	8%
Rest area frequency	492.13 ft (150 m)	49.21 ft (15 m)	41.01 ft (12.5 m)	35.14 ft (10.71 m)	30.77 ft (9.38 m)

* Based on Table 14.10 in New Zealand's *Pedestrian Planning and Design Guide*; amended with Italian-EC law's standards

Table 7 Types of crossing and traffic volumes

Traffic volumes		Pedestrian (peak time)		
		< 100 p/h	100-300 p/h	> 300 p/h
Vehicles (peak time)	< 200 v/h	3	3, 2	3, 2
	200-600 v/h	3, 2	2	2
	> 600 v/h	2, 1*	2, 1*	2, 1*

(*) To use in case of values of street light cycle>120 seconds

*Table 3-1 in *Guidelines for the Design of Pedestrian Crossings*, ACI

- for steps, the internal length is always smaller than the outside
- it is very difficult to provide rest areas of adequate size.

Rest areas are required where the average slope of urban driving exceeds 3 percent. (Table 6)

5.9 Crossing

There are three main types of pedestrian crossings:
1 staggered-level crossings (see also section 7.3 Bridges and underpasses)
2 traffic-light crossings
3 zebra crossings.

The type of pedestrian crossing used depends on two factors in particular: the traffic volume and the road typology. (Table 7)

Type 1 uses similar structures for both pedestrian and bicycle crossings. Type 3 is the most common and potentially hazardous one. It needs to be protected through specific solutions.

5.9.1 Pedestrian islands

A pedestrian island is a part of a road that is demarcated and protected to provide a stopping point for pedestrians while they cross the road.

Its function is to divide and separate the sections of road for the purpose of increasing the safety and protection of pedestrians. The pedestrian island reduces the time spent by pedestrians on the road where conflicts occur between pedestrians and vehicles, and ensures that there is conflict only with a single direction of traffic at a time.

The pedestrian island can also be of great help in determining the duration of traffic-light phases; for example, in the case of long traffic-light cycles (more than 120 seconds), the inclusion of a pedestrian island, with the reduction of the green-light times for pedestrians, may allow a reduction in the duration of the whole cycle, even if the pedestrian is forced to complete crossing in two separate times. The pedestrian island canalizes the traffic flows, and lowers the speed of vehicles in transit, with further safety benefits.

The realization of a pedestrian island is advisable in the case of a pedestrian crossing at a road with more than three lanes, and is mandatory in the case of more than five lanes.

The sizing of the pedestrian island must be proportional to the volume of pedestrian flows; for its design, precise indications are required especially from the point of view of horizontal and vertical signs. Visibility can be signaled by:
- continuous white stripes, of sufficient length and appropriate design, in the part of road paving which precedes the head of the island
- appropriate protrusion from the road surface at its head
- vertical stripes, reflecting yellow and black, at the end of the pedestrian island
- light devices, yellow, in correspondence with the head of the pedestrian island.

The pedestrian area is recommended to be at the same level as the road. It can be expected to rise only at depths greater than 11.18 feet (93.5 meters), when space is required for the proper implementation of two access ramps for strollers and wheelchairs.

The staggered crossing solution is particularly interesting: the two tracts that the pedestrian must cross are longitudinally offset.

The realization of this type of pedestrian crossing, allowing a wide waiting area, is recommended when the pedestrian flows are high. Moreover, in cases of high vehicular and pedestrian flows, it is preferable that railings at least 3.28 feet (1 meter) high are adopted for the protection of pedestrians.

5.9.2 Curb ramps

A ramp is a link between the sidewalk and the road level. Its width must be at least 4.92 feet (1.5 meters) to allow the passage of a wheelchair and a pedestrian. If there are particular constraints, the minimum to consider is 3.28 feet (1 meter).

The slope of the ramp has to be no more than 8 percent. If there are particular geometric constraints, a higher slope value is allowed to a maximum limit of 15 percent (for a maximum height difference of 5.91 inches [15 centimeters]. (Table 8)

Ramps can be classified into two types:
- **Parallel to the travel direction of the vehicles**
There are two options, depending on the size of the sidewalk:
  - The sidewalk is fully lowered.
  - Only the part of the sidewalk adjacent to the pedestrian crossing is lowered, and at least 4.92 feet (1.5 meters) wide.

- **Perpendicular to the travel direction of the vehicles**
In this case, the middle section should be the same level as the road, rising to be the height of the crossing.

5.9.3 Visibility

Both sides of the sidewalk should be free from obstacles that would inhibit visibility. A cone of visibility of at least 16.4 feet by 6.56 feet (5 meters by 2 meters) should be used in areas with high pedestrian and vehicular traffic with more than 200 maneuvers per day.

If the cones of visibility cannot be guaranteed, convex mirrors at the access roads and visual and audio signage for pedestrians must be provided.

5.10 Intersections of paths

The junctions between paths, bike lanes and shared paths can be relatively simple and does not require the same consideration of factors applied to road intersections. Some considerations to be taken into account are:
- a proper sight distance, low gradient, and adjacent areas free from obstacles
- speed control and pedestrian priority in high traffic areas
- no cross intersections that create conflicts
- parapets and rails must be provided where the pedestrian volumes are high.

5.11 Path terminals

The terminal devices are generally placed where a bike path or a common path ends at a road with a crossing. Their purpose is to limit cars' access to the routes, and to warn cyclists that they are approaching a road.

5.12 Public transportation interface

The design of stops requires criteria derived from the best-practice construction:
- good visibility
- stops with names locally recognizable and easily remembered
- good lighting
- stops are kept clear
- the embarkation point is perpendicular to the direction of travel, and signposting is both visual and tactile
- the points of embarkation are clear of street furniture and signage
- variations in the level of the waiting areas and the boarding area is minimal
- the stop has a route map, bus schedules, and information in real time.

At bus stops, any shelters must be designed so that:
- the approaching bus can clearly see the waiting people
- they are properly illuminated
- they are equipped with seats
- they are protected from the rain and wind
- they are resistant to vandalism.

● **Table 8 Design elements of curb ramps**

Element	Key issues	Additional information
Ramp	normal maximum gradient 8 percent (1:12) maximum gradient 12 percent (1:8)	a gradient of 10 percent should only be considered for constrained situations where the vertical rise is less than 5.91 inches (150 mm); a gradient of 12 percent should only be considered for constrained situations where the vertical rise is less than 2.95 inches (75 mm); slopes more than 12 percent are very difficult for the mobility-impaired to negotiate
Ramp	maximum cross fall 2 percent (1:50)	should be consistent across the whole ramp
Ramp	minimum width 3.28 ft (1 m)	4.92 ft (1.5 m) is recommended
Ramp	maximum width: equal to the width of the approaching walking paths	wider ramps are difficult for the vision-impaired to detect
Ramp	tactile paving	tactile paving is necessary to assist people who are blind or have low vision. For more advice, see Guidelines for facilities for blind and vision-impaired pedestrians
Gutter	transition between gutter and ramp	should be smooth with no vertical face; it could happen when the road has been resurfaced
Landing	maximum gradient 2 percent (1:50)	to prevent wheelchair users overbalancing, or accidentally rolling, and to provide a rest area
Landing	maximum crossfall 2 percent (1:50)	to prevent wheelchair users overbalancing, or accidentally rolling, and to provide a rest area
Landing	width: equal to that of the ramp	
Landing	minimum depth 3.94 ft (1.2 m) (top landing)	a depth of 4.92 ft (1.5 m) is preferred

Based on Table 15.2 in New Zealand's *Pedestrian Planning and Design Guide*, amended with Italian-EC law's standards

A quick design guideline for walking paths is summarized below, in Table 9:

● Table 9 Design elements of curb ramps

Topic	Required or recommended
Access for people with disabilities	In general, accessible design requires the elimination of obstacles within the route of travel, 3.28-feet (1-meter) minimum width of travel route, 4.92-feet (1.5-meter) passing areas every 196.85 feet (60 meters) on accessible routes less than 4.92 feet (1.5 meters) in width, maximum grade of 5 percent, steeper grades of up to 5 percent may have ramps and 4.92 feet (1.5 meters) level landing areas for every 2.92 feet (80 centimeters) in elevations change along 8 percent ramps.
Crosswalks	A marked crosswalk includes the use of pavement markings and either signs or signals. Pavement markings should not be used alone to indicate a pedestrian crossing, and signs should be supplemented by pavement markings. Crosswalk signs should not be where pedestrian or full vehicle signals are in place. Stop bars, or twin lines for pedestrian crossings, are suitable only where the approach is controlled by means of a signal or stop sign. Zebra markings are recommended where there are no signal controls as they are more visible to drivers. The length of the zebra stripe differs according to traffic speed 8.20 feet (2.5 meters) minimum where speed is 31.07 miles (50 kilometers) per hour or less, longer where speed is more than 31.07 miles (50 kilometers) per hour.
Special crosswalks	Special crosswalks include pavement markings, internally illuminated overhead signs, down lighting of crosswalk, push buttons, timers, and overhead flashing beacons. These devices can be used in combination to make a crosswalk safer and more effective. Where traffic speeds and volumes are very high, grade separated crossings provide the best protection and ease for crossing pedestrians.
Curbs and edge markings	Curbs are useful to provide a physical separation between pedestrians and traffic. They stop vehicles from mounting the curb for parking, and the gutter acts as a path for stormwater drainage. In rural areas, a curb may seem too urban, and a ditch or swales provide separation. An extruded curb is not recommended where there are bicycle lanes, and may interfere with drainage.
Drainage grates	Drainage grates are best if located outside the route of pedestrian travel; if not possible, the openings should be less than 0.51 inches (13 millimeters) in width and should be mounted flush with the surrounding sidewalk surface.
Hand rails	In steep areas, continuous handrails are to be provided at a height of 3.28 feet (1 meter) to help people in danger of slipping and falling.
Grades	An accessible route of travel should not exceed a grade of 5 percent. If the grade must exceed this maximum, a ramp of not greater than 8 percent may be constructed. Landings of 4.92 feet (1.5 meters) in length are required for every 2.46 feet (75 centimeters) of height, and must have handrails and railings. There are exceptions where the distance is minimal, though a slope of greater than 12 percent is difficult for many users.
Sidewalks	The minimum acceptable width for sidewalks is 4.92 feet (1.5 meters) on local streets and 5.91 feet (1.8 meters) elsewhere; greater width is required where there are greater numbers of pedestrians. Where a walkway is less than 4.92 feet (1.5 meters) wide, passing areas must be installed. Vertical clearance must be a minimum of 6.6 feet (2 meters), but a minimum of 7.87 feet (2.4 meters) is recommended. A cross slope must not be greater than 2 percent but must allow for adequate drainage. Sidewalks must not tilt where driveways cross the street as this creates difficulty people who may be mobility-impaired. There are acceptable designs requiring an extension of a level sidewalk into the driveway; dipping the entire sidewalk where crossed by a driveway may result in drainage problems and add complications to sidewalk travel.
Sidewalk ramps (curb cuts)	Ramps are useful for all people, strollers, luggage wheels, in-line skaters, cyclists, and people in wheelchairs. They provide accessibility at intersections, building entrances, and other areas where elevated walkways are edged with curbing. It is recommended that curb ramps have a detectable warning surface for people who are visually impaired. A warning surface is required at transit ramps. Ramps must be included on two sides of a corner to point pedestrians across to the other curb and must be 3.28 feet (1 meter) wide with a maximum slope of 15 percent. Curb cuts for multi-use paths should be the full width of the pathway.
Street furniture	Street furniture signs, trash cans, and utility boxes may pose hazards to the visually impaired. In general, it is suggested that street furniture be grouped together to increase visibility and take up less room. Add contrast, maintain a clear height of pedestrian walkways, and place grouped objects in an area with a different surface, and/or mark them with a tactile strip.
Street trees	A minimum planting strip is about 6.6 feet (2 meters) wide from the edge of the curb to the edge of the sidewalk. This provides adequate space for the tree to develop, although as little as 3.94 feet (1.2 meters) may be adequate. Tree species should be carefully chosen for good performance.
Tree roots	Potential hazards from tree roots can be controlled by laying a good base of crushed gravel above the tree roots and below concrete sidewalks so they can grow without causing cracks in the sidewalk. Tree roots that may be a hazard to pedestrians can be painted yellow as a warning.
Surface	Smooth surfaces such as cement, concrete or asphalt are firm and stable enough to support wheelchairs, crutches, and other mobility aids. Smoothed gravel screenings may be acceptable in recreational settings, however, loose gravel and wood chips generally do not provide an accessible surface.

Based on Appendix 2, "Quick Facility Design Guidelines" in *Pedestrian and Bicycle Planning. A Guide to Best Practices*; amended with Italian-EC law's standards

6 Design Criteria for Bicycle Paths

6.1 Design speeds

The expected speed of a bicycle route is the first conditioning factor of a project. A design speed of 12.43 miles (20 kilometers) per hour is appropriate for a local path or a path to a main building/zone where there is significant interaction with pedestrians. For other main roads, the designers aim to provide a design speed of more than 18.64 miles (30 kilometers) per hour.

6.2 Widths and clearance required by cyclists

The space required by cyclists in motion must be taken into account:
- "dynamic width" of the cyclist 3.28 feet (1 meter)
- safe distance from fixed objects
- distance from the other modes of traffic (cyclists and motor vehicles)
- high cycle/pedestrian volumes, steep gradients, curves.

The minimum width required is calculated with the equation: minimum width = a+b+c+d, where a refers to dynamic width; b refers to minimum passing distance from other users (Table 10); c refers to clearance for edge constraints (Table 11); and d refers to additional width for high cycle/pedestrian volumes, steep gradients, and curves.

6.3 Gradients

Infrastructure should meet design standards for width, gradient, and surface quality, and cater for all types of user, including children and the disabled. Pedestrians and cyclists benefit from well-maintained and regularly swept surfaces with gentle gradients. Gradient refers to the longitudinal slope along the direction of movement. Steep gradients can deter cycling, but may attract cyclists if the distance is relatively long and requires less effort.
- 3 percent is a desirable maximum gradient.
- 5 percent is the normal maximum gradient, up to a length of 328.08 feet (100 meters).
- 7 percent is a limiting gradient, up to 98.43 feet (30 meters).
- 7 percent or above is a desirable for short distances.

6.4 Drainage

The bike paths should have a crossing slope of 2 percent for drainage. A smooth surface is essential to prevent stagnation of water and ice formation.

When a bike path is in contact with an inclined plane that is not paved or on the side of a pit of an appropriate size, it should be expected to accommodate drainage. A basin can also be provided to facilitate drainage of the area adjacent to the bike path, in view of preserving the existing plants. Any drainage grates and manhole covers should be placed out of the bicycle path.

6.5 Urban bicycle facilities

Paths that can be built between the intersections are:
- curbside bicycle lanes
- bicycle lanes next to parking
- contra-flow bicycle lanes
- bus-bicycle lanes
- paths.

Cyclists do not always need special or dedicated facilities. They need solutions adapted to their needs. For example, curbside bicycle lanes on arteries can have similar benefits to bike paths. However, cyclists prefer to use, where possible, bike paths.

Depending on circumstances, in the absence of dedicated tracks, cyclists can find appropriate solutions through the following means:
- wide curbside lanes
- sealed shoulders
- shared paths
- slow, mixed traffic
- light-traffic streets of adequate width
- unsealed roads and paths
- one-way streets where signs and markings permit two-way use by cyclists.

● Table 10 Minimum passing distance from other users

Speed	Minimum passing distance
Width required for overtaking by motor vehicles	
18.64 miles (30 km) p/h	3.28 ft (1 m)
31.07 miles (50 km) p/h	4.92 ft (1.5 m)
Width required for overtaking cyclist in secondary riding position	
car passing at 18.64 miles (30 km) p/h	14.11 ft (4.3 m)
car passing at 31.07 miles (50 km) p/h	15.75 ft (4.8 m)
bus or heavy goods vehicle passing at 18.64 miles (30 km) p/h	16.73 ft (5.1 m)
bus or heavy goods vehicle passing at 31.07 miles (50 km) p/h	18.37 ft (5.6 m)

Based on Table H.1 in *Sustrans Design Manual: A Handbook for Cycle-Friendly Design*; amended with Italian-EC law's standards

● Table 11 Additional clearances to maintain effective widths for cyclists

Type of edge constraint	Additional width required
flush or near-flush surface (including shallow angled battered curbs)	—
curb up to 5.91 inches (150 mm) high	add 7.87 inches (200 mm)
vertical feature from 5.91 to 6.3 inches (150 to 600 mm) high	add 9.84 inches (250 mm)
vertical feature above 23.62 inches (60 cm) high	add 19.69 inches (500 mm)

Based on Table H.1 in *Sustrans Design Manual: A Handbook for Cycle-Friendly Design*; amended with Italian-EC law's standards

However, it is always necessary to ensure a bicycle route, though differing in various sections, is kept consistent on the whole.

6.5.1 Curbside bicycle lane

A curbside bicycle lane is a bike lane next to a sidewalk for the exclusive use of cyclists. Generally, the signage is constituted by a dotted line and bicycle pictograms at regular intervals.

If there is no need for parking spaces, bike lanes are a good option along roads and should be applied permanently. It is preferable for them to be by the curb instead of next to parked cars.

Advantages of curbside bicycle lanes:
- All road users easily recognize the bike path.
- They provide a fair degree of separation between motorized traffic and cyclists, and highlight the priority lane on the road.
- They can be realized with low costs.

Disadvantages of curbside bicycle lanes:
- They limit the parking area.
- They require constant cleaning to prevent the accumulation of garbage and debris.
- They do not provide sufficient protection for inexperienced cyclists.

6.5.2 Bicycle lane next to parked cars

A bicycle lane can be located next to the parking spaces. Generally, the signage is constituted by one or two dotted lines and bicycle pictograms regularly spaced.

Advantages of bicycle lanes next to parking spaces:
- It eliminates the need to put parking restrictions, giving benefits to other road users.
- It increases the ease of parking as well as the entry and exit of parked vehicles.
- It reduces the distance of travel of pedestrians on the road.
- It improves the funnel traffic, favoring a more orderly and predictable flow of traffic.

Disadvantages of bicycle lanes next to parking spaces:
- A significant carriageway width is required.
- Cyclists could go crashing into a car door as it opens.
- Parking the car could cause inconvenience to cyclists, and potentially cause conflict.

6.5.3 Contra-flow bicycle lane

The contra-flow lanes allow for riding a bicycle against the legal travel direction of a one-way street. They have the same characteristics as traditional bike paths, and are placed so that the cyclists proceed in a normal position on the right (in countries with right-hand drive instead of proceeding on the left) against the flow of vehicular traffic. Contra-flow lanes should be made of materials distinctly different from or with a strong contrast to that of the road, as well as clear markings showing the opposite direction of travel. Any new proposed contra-flow lanes should be well publicized.

An advantage of contra-flow cycle lanes is that they help to give immediacy and consistency to the bicycle path network, allowing cyclists to avoid deviations from the most direct route.

Disadvantages are that other road users, including pedestrians, do not realize that cyclists may be traveling in the opposite direction to the traffic; and contra-flow lanes generally hinder the possibility of parking on the side streets.

6.5.4 Bus lanes

A bus lane is a reserved lane for buses in which cyclists are authorized to travel. Generally, bus lanes can be used by cyclists unless they are specifically excluded by the signs.

The advantage of bus lanes is that buses use these lanes mostly during peak hours, leaving them mostly available for cyclists for the rest of the day, so that the lanes are fully utilized.

The disadvantage of bus lanes is that if the lane is too narrow, the bus will block the path for the following cyclists, forcing them to stop regularly at bus stops; at the same time, buses may be partially slowed by groups of cyclists that precede it. The lanes should be designed with sufficient width to allow the bus to pass cyclists within the lane. (Table 12)

● Table 12 Bus lane widths

Absolute minimum	Minimum	Recommended
10.66 ft (3.25 m)	11.48 ft (3.5 m)	13.12-14.76 ft (4.0-4.5 m)

Based on *Sustrans Design Manual: A Handbook for Cycle-Friendly Design*; amended with Italian-EC law's standards

● Table 13 Widths of exclusive bicycle path

	One way	Two way
Minimum path width	4.92 ft (1.5 m)	8.2 ft (2.5 m)
Absolute minimum	3.28 ft (1 m)	6.26 ft (2 m)

Based on Italian-EC law's standards

6.6 Types and widths of bicycle paths

There are mainly three types of bicycle paths, each of which can be separate from the streets or alongside them.
- exclusive bicycle paths
- unsegregated shared-use paths
- segregated shared-use paths.

6.6.1 Exclusive bicycle paths

An exclusive bike path can be used only by cyclists, and it is preferable where a significant volume of commuter cyclists is expected. There are two kinds: one-way bike paths and two-way bike paths. (Table 13)

Advantages of an exclusive bike path:
- Cyclists can proceed normally with no delays caused by conflicts with other users.
- Cyclists have a higher level of safety and better riding experiences.

A disadvantage is that city ramblers sometimes use the exclusive bike paths when their facilities are relatively poor.

6.6.2 Unsegregated shared-use path

A shared path is simultaneously used by cyclists and pedestrians. It is suitable where they both need a path, but their numbers are modest. (Figure 13)

For shared paths, particular attention must be paid:
- where cyclists join the shared path; it has to be ensured that they can do it safely without conflicts with pedestrians
- where the shared path ends, the cyclists do not continue to use the path reserved for pedestrians
- where it crosses another pedestrian, bicycle, or shared path
- to providing adequate visibility to cyclists who have higher speeds than pedestrians
- to providing appropriate signage to indicate the presence of both pedestrians and cyclists.

It is important that:
- the design of a path is compatible with demand and usage requirements (Table 14)
- local authorities can monitor users' behavior
- connections between path, road, and driveways should be carefully considered.

Avantages of shared-use paths:
- They are useful for cyclists and pedestrians, and therefore maximize the benefits to the whole "slow-mobility community."
- They require low-cost interventions, because existing path can be adapted or extended.

Disavantages of shared-use paths:
- The conflict between cyclists and pedestrians is evident in cases where, for example, there is a significant volume of them or where interference will be created between the pedestrian needs and those of the cyclists.

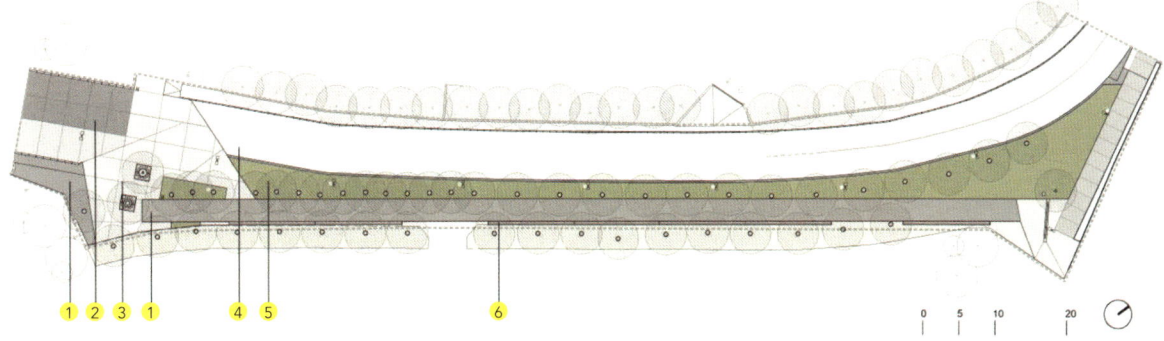

Figure 13 Unsegregated shared path

1. Sidewalk
2. Main road
3. Pedestrian crossing
4. Internal street
5. Buffer zone
6. Linear seat

Table 14 Widths of unsegregated shared-use paths

	Local access only	Recreational or mixed use
Minimum path width	8.2 ft (2.5 m)	13.12 ft (4 m)

Based on Italian-EC law's standards

Table 15 Widths of segregated shared-use paths

	Area for cycles	Area for pedestrians	Total
Minimum path width (two-way)	6.56 ft (2 m)	4.92 ft (1.5 m)	11.48 ft (3.5 m)
Absolute minimum (one-way)	3.28 ft (1 m)	4.92 ft (1.5 m)	8.2 ft (2.5 m)

Based on Italian-EC law's standards

- This solution provides a significantly reduced cycling standard, which causes lower speeds.

6.6.3 Segregated shared-use path

A segregated shared-use path is a path with separate sections for cyclists and pedestrians, and is appropriate when a large number of users is expected. Pedestrians and cyclists need to use segregated areas of the path outlined by surfaces differentiated through material or color (Figure 14). The separation can also be guaranteed by:

- an insurmountable raised curb
- a horizontal thermoplastic signage
- a plant barrier
- a pedestrian area of at least 4.92 feet (1.5 meters). (Table 15)

Advantages of segregated shared-use paths:
- Separate paths avoid conflict between pedestrians and cyclists.
- Cyclists do not experience delays due to the presence of pedestrians.

Disadvantages of segregated shared-use paths:
- Higher speeds are technically possible, but the possibility that pedestrians can cross over into the bike path leads to a higher perception of accident-risk.
- Separate paths are wider than the other routes, and are more expensive.

6.7 Signing and marking

Adequate signage is essential on a bicycle path, especially to warn cyclists of potential conflicts and to communicate clearly the regulations of the crossings. It's necessary to report any design elements that deviate from the minimum-size criteria (for example, bending radii, horizontal or vertical distances lower than the minimum prescribed or recommended). In addition, the implementation of the reporting practices established by regulatory standards, or by guides (for example, directions, locations, distances, route numbers, names of streets crossing) is used in the same way as in the streets. It is particularly useful in the following circumstances:
- near roads with particularly heavy traffic
- on curves with limited visibility
- on dark paths and in dark areas.

Figure 14 Segregated shared-use path

Table 16 Quick facility design guidelines: bicycle path

Topic	Required and recommended
Bicycle parking	Secure short-term and long-term parking must be provided at all destinations; parking can be required; careful design and placement criteria are recommended for best results.
Bridges and over-passes	Special attention is needed to ensure adequate protection from traffic, adequate railing height and materials, and adequate width for sharing with pedestrians. A railing at handle-bar height and one at shoulder height should be provided. Vertical railings or chain-link fences that can easily snag a handle bar and cause a crash must be avoided.
Construction zones	Bicycle lanes are to be rerouted for construction; adequate signage and lighting must be in place. Where metal plates provide temporary road surfaces, they must meet the road at right angles, and a ramp of asphalt provides a feathered edge for cyclists.
Extruded curbs	Extruded curbs should not be used to separate a bike lane from traffic as they present a hazard to the safe operation of a bicycle, making left-hand turns impossible and causing cleaning difficulties.
Drainage/utility covers	Drainage and utility grates should be flush with the road, and long openings should be placed at right angles to the direction of travel. Ideally, grates and utility covers should not be placed in the bike lane, and curb inlets should be used instead.

Topic	Required and recommended
Intersections	Intersections (including driveways) are the most likely places for car–bike collisions. Intersections should be carefully designed to reduce the chance of conflict. Driveways should have adequate sight lines to see all traffic on the road. Bike lanes at intersections and bike paths where they connect with streets should be carefully designed. Intersections with freeways should be grade-separated.
Lighting	Bicycle facilities should be adequately lit. Street lighting is usually sufficient for wide curb lanes and bike lanes; separated paths and bike parking areas require appropriate lighting where evening walking and cycling is expected. Intersections of paths with roads must be well lit.
Maintenance	Regular maintenance is essential to ensure that the facilities are safe and comfortable to use. Road and path surfaces should be swept regularly to remove glass and other debris. They should be given the same or greater maintenance standard as roads due to the absence of the "sweeping action" from regular car travel.
On-road facilities: bicycle lanes	All roads should be thought of as forming the bicycle network. On major urban roads, bike lanes can increase safety and reduce conflicts with other vehicles. Bike lanes should always be one-way, in the same direction as adjacent traffic. The absolute minimum width for a one-way bike lane should be 3.28 feet (1 meter) excluding curb and gutter, and 4.92 feet (1.5 meters) when next to a parking lane. Bike lanes should end well in advance of intersections, with dashed lines adjacent to turn lanes to encourage traffic to merge into the bike lane before turning. Bike lanes should be located to the far right of the road, or between the parking lane and the travel lane.
Bicycle boulevards and local streets	Bicycle boulevards are streets that encourage cycling and discourage motor vehicles by means of traffic-calming devices. On local streets, bicycle-route signs may be desired where they form part of a bicycle network.
On-street parking	On-street parking can pose risks to cyclists who ride past and people as they exit their vehicles. Where cars are parallel parked, a bike lane may be provided between the road and parked cars if the bike lane is wide and far enough from the vehicles to avoid car doors opening into the bike lane. A bike lane should never be placed at the right of parallel parking. Diagonal or perpendicular parking is very dangerous, and bike lane should be avoided in these areas.
Pavement structure	A bike lane may be cement, asphalt, or fine gravel screenings. However, the surface should be at least as smooth as that provided for vehicles, and tree roots should be prevented from disrupting the smooth surface.
Railroad crossings	Railroad crossings should be at right angles to the rails, otherwise they may trap the wheels and cause crashes. The travelway should be widened if the crossing angle is less than 45 degrees to permit a wider crossing angle. Warning signs and pavement markings should be posted before the crossing. Road surfaces should be flush with the rails. Rubberized flanges around the rails or removal of unused track can minimize the danger for cyclists.
Intermodal linkages	Airports, rail, buses, and ferries permit cyclists to reach distant destinations. All trains should be designed to permit bicycles as checked baggage, or in the passenger car. Terminals should provide secure bicycle parking, and areas may be provided for bicycle setup, and clear access to the station should be provided. Transit buses should be equipped with racks to carry at least two bicycles.
Ferries	Ferries sometimes represent a vital link in the transportation system. Provide for bicycles on vehicle and passenger ferries and at ferry terminals by providing dedicated bicycle routes through the terminal to boarding areas, and secure and protected parking at the terminal and on the ferry to prevent damage, theft, and weather exposure.
Separated paths	Separated bicycle paths should not be thought of as a substitute for accommodating bicycles on nearby roads. These paths should be considered as extensions of the street system and meet an important recreational need. Two-way paths need to be carefully designed where they intersect with traffic. Twinned paths on each side of a road provide more safety, especially at intersections. The minimum width for a two-way path is 8.2 feet (2.5 meters) and a wider (more than 13.12 feet [4 meters]) path with markings down the center of the path may minimize conflicts where there is heavy traffic. Rail lines converted to trails have good sight lines and shallow grades. Good access, motor vehicle parking, water, toilets, and telephones make for a successful trail.
Sidewalks and ramps	Cycling on the sidewalk is generally not recommended for safety reasons, as there is a high potential for collisions at driveways and intersections.
Traffic control devices	As bicycles are legally able to travel on the roads, they do not require special traffic control devices. The same standards that apply to street signs and highways also apply to bicycle tracks. High-traction, non-skid paint should be used on road surfaces.
Traffic control devices bollards	Steel bollards (or painted and with reflectors) should be placed where vehicles may enter a bike path; one should be placed in the center, with another on each side, each providing 4.92 feet (1.5 meters) clearance.
Traffic signs	Standard signs are adequate for most bicycle paths. Signs specifically directed at cyclists should be smaller and lower than normal street signs. Signs should be between 3.94 feet (1.2 meters) and 9.84 feet (3 meters) in height and should be 3.28 feet (1 meter) from the edge of the bicycle path to provide adequate clearance for cyclists who may veer off the path to pass. Consideration should be given to adequate stopping distance to heed the warning or information on the sign. Bike route signs should be used in conjunction with subplates indicating destinations (with distances) to be found along the signed route. In addition, bike route signs must be part of a comprehensive system. At junctions of separated trails with roadways, the name of the road should be clearly visible to trail users.
Traffic-calming devices	Traffic-calming measures usually benefit cyclists by removing or slowing traffic. Some measures need to be carefully designed to accommodate cyclists. For example, where speed bumps or diverters are used, a bypass area for cyclists should be included. Where pinch-points are used, rolled curbs reduce the danger of being squeezed. Traffic-calming devices can also be used as refuges for cyclists crossing two-way busy roads. Refuges should be 10 feet by 6.6 feet (3 meters by 2 meters), and provide handrails and bollards.
User conflict	Design features and user policies should be used to minimize conflicts between cyclists and pedestrians.
Vegetation	It is important that vegetation near roadways and paths be maintained. All vegetation taller than 10 feet (3 meters) should be trimmed back at least 3.28 feet (1 meter) on each side of all paths. Vegetation at intersections should be trimmed to provide adequate sight lines. Tree and shrub roots may cause disruption in a path surface, removal of trees within 3.28 feet (1 meter) of the path and the use of root barriers may help to reduce problems.
Workplace facilities	Many people say that they would try commuting by bike, but feel they need a shower and a place to change clothes once they arrive at work. Some jurisdictions are requiring that such facilities be provided when a building is built or remodeled. Clothes lockers, large enough to accommodate a week's worth of clothes and toilet articles, can be provided. A bathroom may be remodeled to add a shower stall.

Based on Appendix 2: "Quick Facility Design Guidelines" in: *Pedestrian and Bicycle Planning. A Guide to Best Practices*; amended with Italian-EC law's standards

7 Intersections of Paths and Roads

The key planning principles relate to the type of intersection control used and the provision of adequate space:
- The intersection should perform efficiently for cyclists under the different traffic conditions expected throughout the planning period.
- It must be suitable for cyclists of basic competence.
- All normal maneuvres should be possible, particularly left turns (including the option of hook turns).
- The conflict area between through-cyclists and right-turning traffic (especially heavy vehicles) needs to be managed.

Right-turn slip lanes can simplify this by moving right-turning traffic conflict points away from the intersection and providing space for hook turns. It also means:
- conflict points are easily identified
- cyclists and drivers know where cyclists are expected to be on the road
- the intersection is consistent in alignment and standards with mid-block facilities on approach and departure.

7.1 Priority at sideroads (meeting secondary streets)

The connections with the side streets are crucial. Any bicycle path along a road should be as continuous as possible: in other words, the cycling priority should be maintained on any secondary road.

Frequent stopping makes cycling much harder. The priority for the bicycle lanes works best when the path is uni-directional (cyclists going in the same direction as other vehicles).

The sidewalk adjacent should ensure a continuous path for pedestrians, thus strengthening the priority of the non-motorized travelers coming from the side streets. To emphasize this priority, crossings can be highlighted by a red color. The bicycle path and sidewalk should have to change height when crossing the secondary street.

At intersections, it is good to provide a small waiting area for motor vehicles, so that vehicles that are to engage the crossing do not block the cyclists. For hybrid bike paths, simply continue the lane through the junction, keeping the red pavement.

7.2 Junctions of primary and secondary streets

Collisions are more frequent at intersections. Minimizing the points of conflict can reduce the chance of accidents, creating a greater perception of security that could increase the number of cyclists.

Simple intersections may be used for secondary roads with low traffic volumes. The basic requirements for safe, convenient, easy-to-use crossings for cyclists are:
- maintain the separation of ways through the intersection to reduce the number of interactions with other traffic
- ensure visibility, using perpendicular crossing points
- equip the busy streets with waiting places
- avoid the need for lots of stopping
- reduce the speed of traffic through junctions.

7.2.1 Maintaining separation through the junction

The space dedicated to cyclists must continue through the intersection so that cyclists can benefit from a continuous path that avoids interaction with traffic. The bicycle path must be marked through the junction to indicate the cyclists' space and to reduce the risk of collisions with vehicles turning right.

7.2.2 Visibility

Key principles in the design of a public space are efficiency and safety, based on the fact that a user can't look in several directions at once, to guarantee maximum security when negotiating an intersection.

This can be achieved by the combination of:
- protective islands
- breakpoints of bicycle paths
- perpendicular crossings.

7.2.3 Providing bicycle refuges on large primary streets

For maximum safety, bicycle paths should never cross more than one lane of traffic without a safe haven where cyclists can wait. Shelter islands should be provided in these lanes, at a length of at least 7.87 feet (2.4 meters). If it is not possible to provide an acceptable safety level, there should be at least a sign to highlight the separation of the bike path from the road. (Figure 15)

7.2.4 Avoiding the need for lots of stopping

Frequent stops make cycling very difficult and unappealing. Tunnels or bridges could be used to maintain a continuous path, even where it is not possible to attribute priority. Tunnels are preferred, but only if the view through the tunnel allows the user to see the end before entering. Bridges should be as straight as possible with gentle ramps. In the case of new projects, the architect should consider raising the road to reduce the depth of an underpass, or consider lowering the road to reduce the height of the bridge.

7.2.5 Reducing the speed of traffic through junctions

Signalized intersections should ensure reduced waiting time for bicycles, with short but more frequent traffic-signal cycles. Waiting for more than 30 seconds will make the use of bicycles less attractive. The cyclists in bike lanes must have the same options of negotiating the intersections as car traffic. Consideration should be given to using simultaneous green-light phases for bicycles.

7.3 Bridges and underpasses

An overpass, an underpass or a small bridge may be necessary to give continuity to a bicycle lane. Their minimum clear width should be the same as the bike path, but the width recommended range should be increased by at least 23.62 inches (60 centimeters) on each side.

The increasing of the clear width in correspondence with a path structure has two advantages. First, it provides a minimum horizontal distance from the parapet or barrier; and second, it provides the room for the necessary maneuvers to avoid conflicts between pedestrians and other cyclists who are stopping on the structure. For example, the extra width can be justified at particularly attractive places (over a river or ravine), where users stop to enjoy the view.

The bridges designed exclusively for pedestrian traffic can be designed for pedestrian loads, but in general, for security and servicing, they must be designed to bear greater loads. (Figure 16)

The railings and parapets on bridges must have a minimum height of 43.31 inches (110 centimeters). In cases of areas affected by occasional water flow, as an alternative to bridges, paved fords can be considered.

The tunnel design should follow the criteria of size and clearances previously determined, as well as the preliminary identification of the other types of traffic that might use the structure (for example, emergency vehicles). In tunnels longer than 98 feet (30 meters), the sense

Figure 15 Cycle refuge

Pottery Road bicycle and pedestrian crossing—PLANT Architects Inc, Photo by Peter Legris

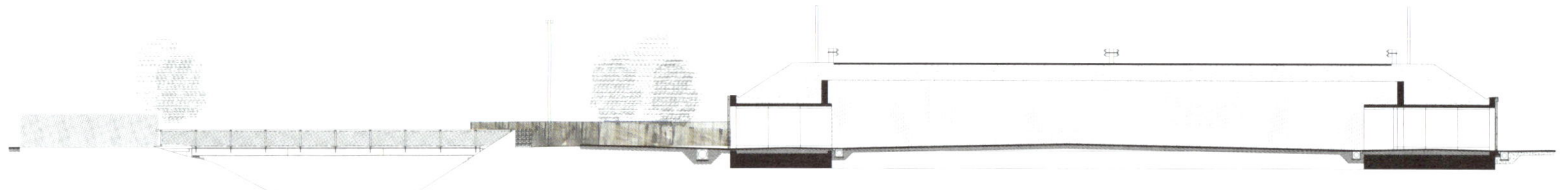

Figure 16 Bridges

of security of the users can be improved by larger entrances or internal sections (minimum 10 feet [3 meters] high and 13.12 feet [4 meters] wide). The path should provide a clear view through the entire length of the tunnel. The lighting (if possible with zenithal opening to ensure ventilation) should be especially considered in areas where security is a priority. The drainage of overlying road infrastructure must be routed in such a way as not to affect the surface of the path.

8 Provision Parking Facilities

Parking spaces for bicycles are essential in a multi-modal transportation system. In fact, in addition to helping to prevent theft, the installation of well-designed parking in appropriate locations can make the sidewalks clearer and more attractive. In the absence of bicycle parking or parking lots, which are particularly comfortable, people tend to secure their bicycles to any fixed object, signs, parking meters, fences, or trees. These bikes can be placed randomly, obstructing pedestrians or vehicular flows, and making a sidewalk inaccessible to people with disabilities. Integrated parking (with cars and motorcycles) can be economically advantageous, increasing the overall supply and the resulting profitability of parking.

Bicycle parking should be provided at both origins and destinations. Two main categories of parking are short- and long-term. The needs of each differ in terms of the design and level of protection. In many places, it may be appropriate to include a combination of short- and long-term parking facilities. (Figure 17)

8.1 Planning for bicycle parking

Parking spaces for bicycles can be designed and installed in various ways, but should primarily be placed at all nodes of modal interchange at public facilities. It should be an integral aspect of the development process and rebuilding. The involvement of local cycling associations in the planning process can be a great way to determine the best locations for parking spaces.

The construction of bicycle parking can be required in new developments and urban regeneration. It is a cost-effective way to equip cities with bicycle parking. Another approach is to require the provision of a certain number of parking spaces, although this requirement contradicts the efforts to reduce the parking of motor vehicles and increase the proportion of walking and cycling. The need for bicycle parking may increase over time, and planning should anticipate this possibility.

Cyclists always try to park as close as possible to their destination. The parking spaces for bicycles should therefore be conveniently positioned in a manner that is clearly visible and the nearest to the building entrance of the destination.

8.2 Directional signage for bicycle parking

The bicycle racks should be positioned so they:
- are easily accessible from the street and protected from motor vehicles
- are visible to passers-by to promote its use and improve safety
- do not obstruct or interfere with pedestrian traffic or maintenance activities
- do not prevent access to buildings, public transportation, and loading/unloading areas
- leave a reasonable space for opening the doors of parked cars
- are covered, if possible.

8.3 Short-term bicycle parking facilities

Short-term parking facilities should be provided wherever people may need to leave bikes unattended for a short period of time. In general, bicycle parking should be considered wherever there is car parking, and in high-density places, such as the downtown areas.

Bicycle parking should be easy to spot and use. It is best placed mainly in shops, restaurants and bars, apartment buildings and condominium complexes, offices and public facilities, such as public transportation stations, schools, parks, and libraries. Two key components of short-term parking spaces are location and accessibility.

Figure 17 Parking facilities

Park[e]ing—Stradivarie Architetti Associati, photo by Gina Omenetto

9 Maintenance and Management

Maintenance and management are critical to the successful operation of walking and cycling infrastructure. A path maintained in good condition is usually more popular than the one left to deteriorate, for it provides a safer and more comfortable walking or cycling experience. Regular maintenance helps protect a path from deterioration, reducing the need for expensive rebuilding later. It is essential for the project design team to consider future maintenance arrangements at the start of the development phase, especially when the funding for maintenance has to be factored into existing budgets. A high-quality, suitable design will mean less ongoing maintenance in the future and a better quality of space. (Figure 18)

Figure 18 Signage and lighting

Adolf B. Horn Avenue—AGRAZ Arquitectos

A good maintenance plan will cover at least the following:

- ### Surface maintenance
  Pedestrians and cyclists are sensitive to the quality of the path surface. It is also necessary to examine the paths and the curb regularly for surface irregularities, such as manhole covers, grilles, pot holes, pavement gaps, or ridges. Such hazards must be repaired quickly. Some guidelines suggest that maintenance of the central 4.92 to 6.56 feet (1.5 to 2 meters) area of a path should be prioritized, since this part is used most frequently by the cyclists or pedestrians.

- ### Sweeping and cleaning
  When establishing a street-sweeping plan, the pedestrian and bicycle routes should be prioritized, as the sand, gravel, fallen leaves in autumn, and snow in winter are more likely to accumulate on the sidewalk or at the road curb. Drainage channels and culverts also need to be cared for. A blocked culvert can result in major soil saturation and a land slip.

- ### Signage and markings
  Path signage and markings degrade or become damaged over time, so they need to be inspected regularly.

- ### Vegetation
  Vegetation growing into the paths must be cut back at least twice a year (during spring and autumn) to prevent leaves and branches from impeding the sight lines, and roots from breaking up the path surface.

- ### Lighting system
  Street lights need to be checked regularly, and damaged street lights must be repaired promptly.

Information about respectful behavior on shared paths is useful for local users, who can also be encouraged to report any damage and obstructions.

Bibliography

ACI, *Guidelines for the Design of Pedestrian Crossings*, Automobile Club Italia, Rome, Italy (2011).

Austroads, *Austroads Guide to Road Design. Part 6A: Pedestrian and Cyclist Paths*, Austroads, Sydney, Australia (2009).

Austroads, *Cycling Aspects of Austroads Guides*, Austroads, Sydney, Australia (2011).

Benjamin, Walter, *Berlin Childhood around 1900*, translated by Howard Eiland, Harvard University Press, Cambridge, MA, United States (2006). The first book publication of Berliner Kindheit um Neunzehnhundert was arranged by Theodor W. Adorno in 1950.

Biton, Anna, David Daddio and James Andrew, *Statewide Pedestrian and Bicycle Planning Handbook*, U.S. Department of Transportation, Federal Highway Administration Office of Planning, Washington, DC, United States (2014).

Cambridge Cycling Campaign, *Making Space for Cycling: A Guide for New Developments and Street Renewals,* 2nd edn, Cyclenation, Cambridge, United Kingdom (2014).

Commission Expert Group on Transport and Environment, *Defining an Environmentally Sustainable Transport System*, Working Group I, Bruxelles, Belgium (2000), http://www.ocs.polito.it/biblioteca/mobilita/Defining.pdf.

Cozzi, Mauro, Silvia Ghiacci and Marco Passigato, *Cycle Paths: Manual and Guide for Traffic Calming*, Il Sole 24 Ore, Milano, Italy (1999).

DT, *Planning and Designing for Pedestrians: Guidelines*, Department of Transport, Perth, Australia (2010–14), http://www.transport.wa.gov.au/activetransport/24033.asp.

Dufour, Dirk, *PRESTO Cycling Policy Guide Infrastructure*, Ligtermoet & Partners (2010), http://www.presto-cycling.eu.

Federal Highway Administration, *Separated Bike Lane Planning and Design Guide*, United States (2015), http://www.fhwa.dot.gov/environment/bicycle_pedestrian/publications/separated_bikelane_pdf.

Federal Highway Administration, *Designing Sidewalks and Trails for Access*, United States (1999), http://www.fhwa.dot.gov/environment/bicycle_pedestrian/publications/sidewalks.

FIAB, *Cycling Networks in the Mediterranean Area: A Handbook of Cycling*, Federazione Italiana Amici Della Bicicletta, Italy (2008).

Gehl, Jan, *Life Between Buildings: Using Public Space*, translated by Jo Koch, Van Nostrand Reinhold, New York, NY, United States (1987).

IDT, *Bureau of Design and Environment Manual*, Illinois Department of Transportation, Division of Highway, Illinois, United States (October 2010).

Litman, Todd et al, *Pedestrian and Bicycle Planning. A Guide to Best Practices*, Victoria Transport Policy Institute, Victoria, Canada (2011).

LTSA, *Cycle Network and Route Planning Guide*, Land Transport Safety Authority, Wellington, New Zealand (2004), https://www.nzta.govt.nz/walking-cycling-and-public-transport/.

MoL, *London Cycling Design Standards*, Mayor of London, London, UK (2014), https://tfl.gov.uk/corporate/publications-and-reports/streets-toolkit.

MoT, *Cycling in the Netherlands*, Ministry of Transport, Public Works and Water Management, Directorate-General for Passenger Transport, Den Haag, The Netherlands (2009).

NATCO, *Urban Street Design Guide*, Island Press, Washington, DC, United States (2013), http://www.nacto.org/publication/urban-street-design-guide/.

NATCO, *Urban Bikeway Design Guide*, Island Press, Washington, DC, United States (2014), http://www.nacto.org/publication/urban-bikeway-design-guide/.

NTA, *National Cycle Manual*, National Transport Authority, Dublin, Ireland (2011), http://www.cyclemanual.ie.

NZTA, *Pedestrian Planning and Design Guide*, New Zealand Transport Agency, Wellington, New Zealand (2009), https://www.nzta.govt.nz/resources/pedestrian-planning-guide/.

NZTA, *Guidelines for Facilities for Blind and Vision Impaired Pedestrians*, Wellington, New Zealand (2015), http://www.nzta.govt.nz/assets/resources/road-traffic-standards/docs/rts-14.pdf

RTA, *NSW Bicycle Guidelines*, Roads and Traffic Authority, Sydney, Australia (2005), http://www.rms.nsw.gov.au/business-industry/partners-suppliers/documents/technical-manuals/nswbicyclev12aa_i.pdf, http://www.rms.nsw.gov.au/.

Sustrans, *Sustrans Design Manual: A Handbook for Cycle-Friendly Design*, Bristol, UK (April 2014), http://www.sustrans.org.uk/our-services/infrastructure/route-design-resources/documents-and-drawings/key-reference-documents.

Welch, Aaron, Kaid Benfield and Matt Raimi, *How to Tell if Development is Smart and Green*, U.S. Green Building Council, Washington, DC, United States (2011), http://www.usgbc.org.

Revitalisation of Spikeri Square and Daugava Waterfront Promenade

Master plan

- Location / **Riga, Latvia** ● Area / **14.08 acres (5.7 hectares)**
- Length / **1 mile (1.6 kilometers)** ● Completion / **2013** ● Landscape design / **Arplan, Ltd., A plus Architects, Ltd.** ● Photographers / **Aivars Silins and Ilze Denisova** ● Client / **Riga City Council City Development Department** ● Budget / **€7,831,152**

Based on the consideration of its geographical closeness to the old city, the revitalization of the degraded Spikeri area has organically improved the urban infrastructure for people and creative industries, services and facilities located in this area. This is done by creating a new and modern open public space that offers new recreation possibilities, reconstructing the former pedestrian underpass, and connecting it to the city with pathways, quays, and a new bicycle lane. The territory now meets all the accessibility requirements. It invites people of all ages to use the open-air public space for a wide range of activities.

To avoid intense traffic noise, green islands are created along Krasta Street (in place of the former warehouses) with a green slope towards the street and seats for the visitors towards the Spikeri Square. The main street in the Spikeri area connects the Central Market and the Daugava Waterfront.

New ramps and stairs lead people from the Spikeri area toward the reconstructed pedestrian underpass, connecting the Daugava Waterfront Promenade with the city and providing access for the mobility-impaired.

The Daugava Waterfront stretches 1 mile (1.6 kilometers) along the Krasta Street and the river, and offers new recreation areas with water-play facilities, quays, sunbathing benches, a new bicycle lane and a skate park.

The waterfront along the Spikeri area extends the promenade in front of the Old Town, offering a wide range of public activities in the open cityscape, as well as exclusive views to Riga's bridges, the Old Town and the new National Library of Latvia. The revitalization of the waterfront includes completely new infrastructure for the embankment that has also been renovated and reinforced.

Local materials and techniques have been used in the creation of the project. Modular benches, green walls and other elements were designed exclusively for this project. The new waterfront infrastructure provides rainwater drainage and lighting. A stormwater management system is connected to the city's sewer system, and new quays for water transportation have been created. New greenery, decorative flower beds and trees have been established.

01 New quays for water transportation
02 Recreation area of the waterfront promenade and railway bridge in the background

03 Recreation area with modular benches
04 Waterfront promenade with Salu bridge in the background
05 Recreation area and the view to the other side of the river
06 Waterfront promenade along Krasta Street

05

06

07 Skate park, sunbathing benches and view from the south-east
08 Terraced concrete skate park
09 New trees with tree protectors
10 Overhung glass view platform
11 Pedestrian underpass interior

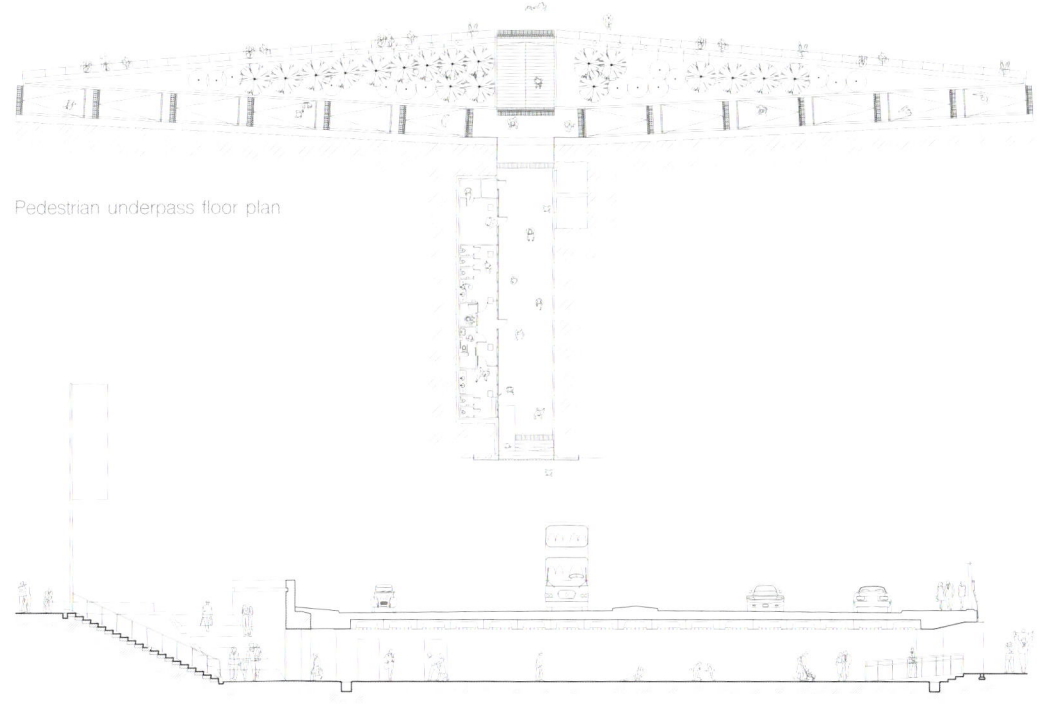

Pedestrian underpass floor plan

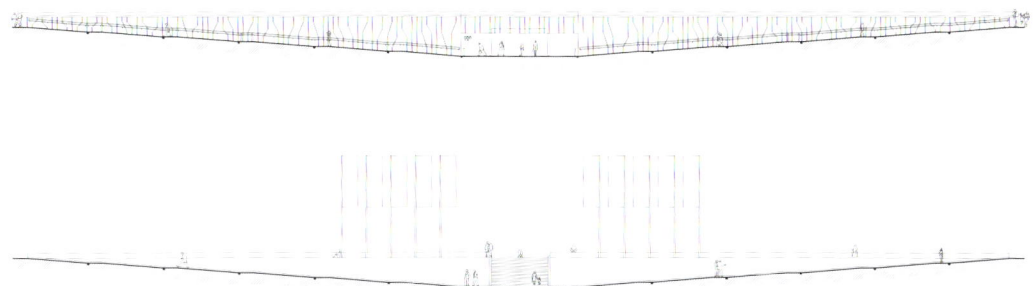

Pedestrian underpass section

Pedestrian underpass elevation

12 Pedestrian underpass entrance from the side of Spikeri Square
13 Green slope and ramps toward the pedestrian underpass
14 Spikeri Square with playground and acoustic object
15 Inner street connecting the square with the Central Market

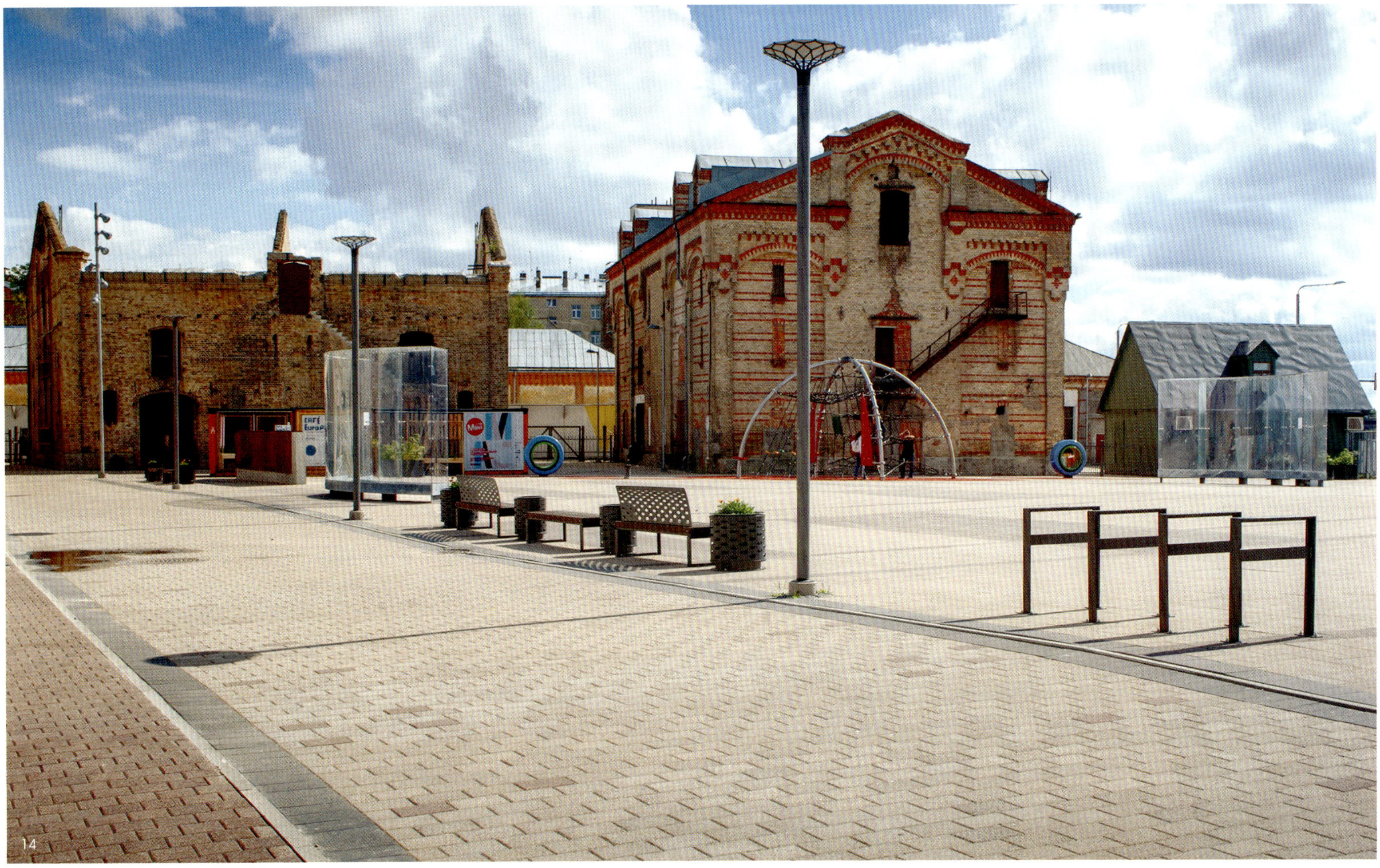

14

Square pavement detail

15

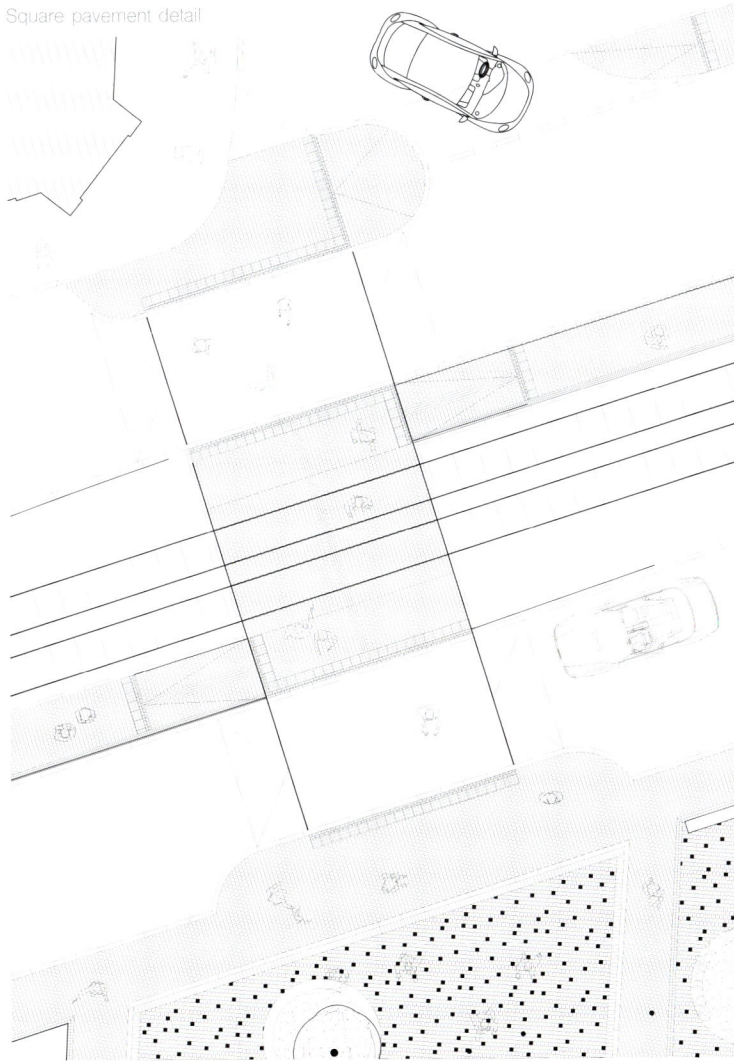

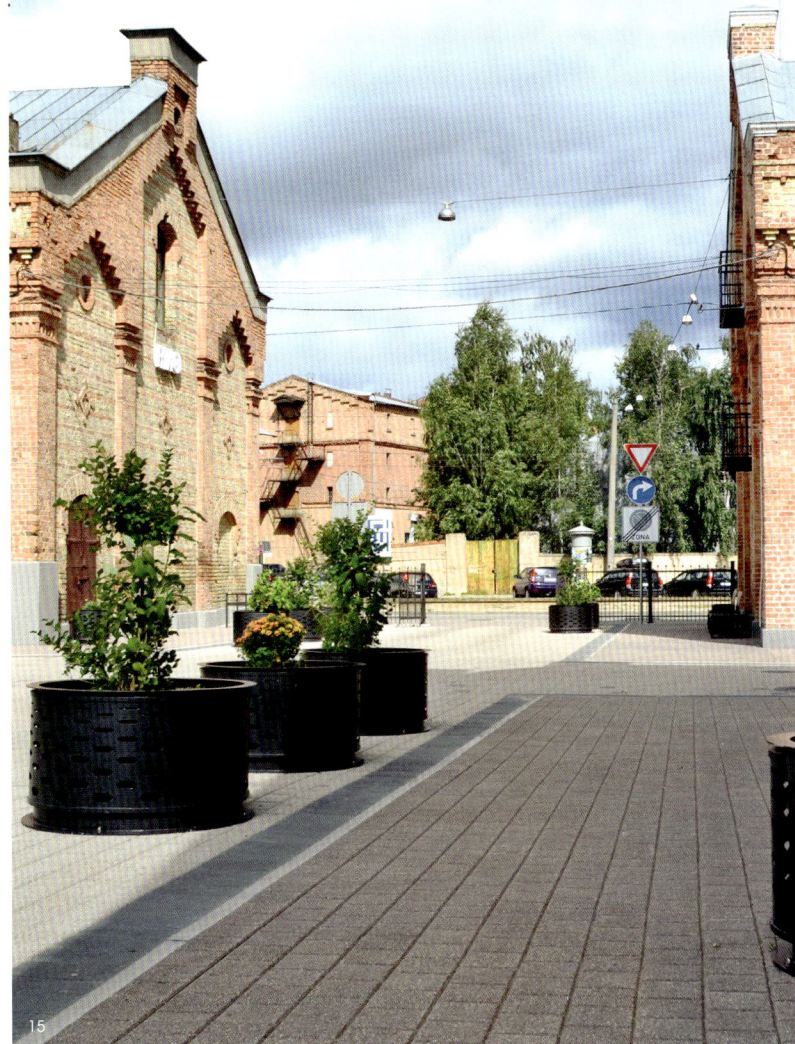

Mapocho 42K Riparian Promenade

1. Providencia
2. Quinta normal
3. Cerro Navia

The promenade runs through eleven districts with different social and economic status (only three are indicated here).

● Location / Santiago, Chile ● Area / 175.44 acres (71 hectares) ● Length / 26.09 miles (42 kilometers) ● Completion / 2015 (1st stage) ● Landscape design / M42KStudio ● Photographer / Francisco Croxatto V ● Client / Housing and Urban Development Ministry, Chilean Government ● Budget / US$6,500,000

The project was initiated by the idea of recovering and consolidating the banks of Mapocho River as a unique public metropolitan space of 26.09 miles (42 kilometers): an east–west backbone of Santiago that connects eleven boroughs with different social and topographical conditions.

The difficulty in recognizing the river as a living landscape of the city is due, in part, to the impossibility of being able to include the entire river. Therefore, Mapocho 42K proposed a geographical and a social continuity: a promenade for pedestrians and cyclists that connects all the existing parks and potential green areas in the vicinity, contributing to a better quality of urban life.

Two concepts are the key to the spatial definition of this project: to consolidate the promenade as an urban balcony along the river, and at the same time, as a green corridor—a tree-covered backbone. It is under the shade of these trees that the walking path and the bike track are to be located. The proposed vegetation and trees for the corridor are mainly species that are adapted to the local semi-arid climate, with low water requirements, such as soapbark trees, and pepper trees. The borders are highlighted with textures and colors that boost the visual appeal of the green corridor.

01 Main access area

1 Rural section (6.21 miles/10 kilometers)
2 West section (6.84 miles/11 kilometers)
3 Central section (4.97 miles/8 kilometers)
4 East section (8.08 miles/13 kilometers)

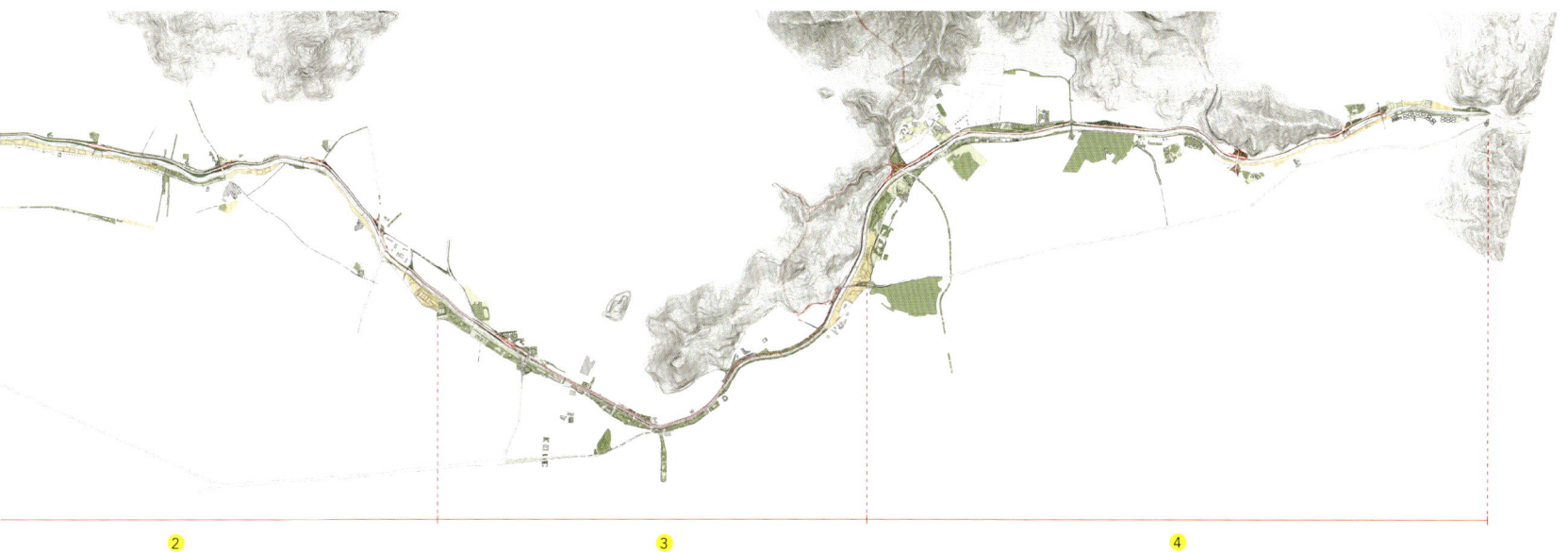

A geographic corridor along the river

West section of the urban balcony project along the riverbank

East section and riverbed

The project is planned as an open design: a structural corridor with a width of 49.21 feet (15 meters) to 82 feet (25 meters) along the riverside, with diverse rest areas, play zones and squares, which may be expanded into adjacent areas in time. At the same time, the design strategy is based on elements and components, which could be implemented by eleven municipalities of different socioeconomic situations to establish a common identity throughout the corridor. This includes criteria such as environmental matters, vandalism, and long-term maintenance. So the project utilizes materials such as colored asphalt, gravel, and prefabricated concreted pieces. The signage system emphasizes an accessible bike path within a park environment.

02 Promenade as social and geographic continuity

Cerro Navia riverbank plan

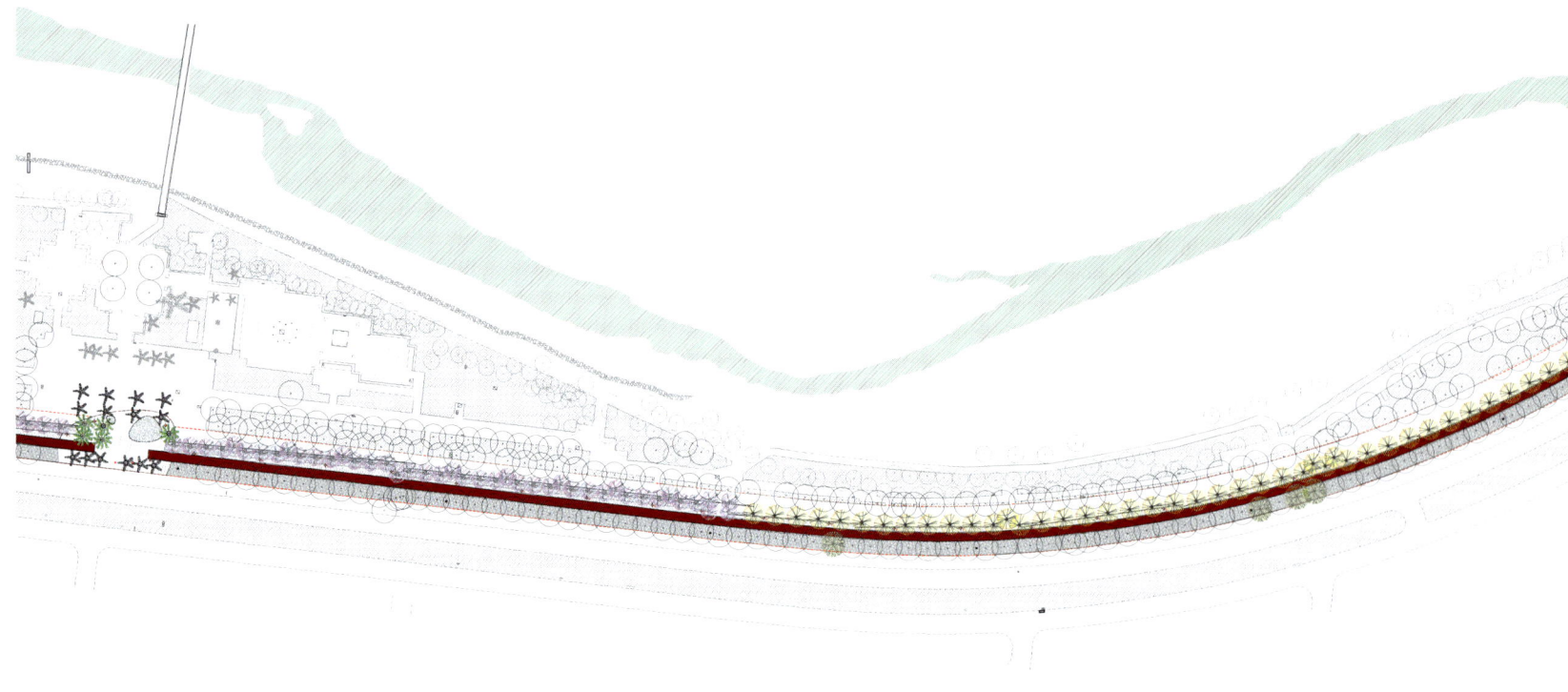

Quinta normal riverbank plan

1. Bike path
2. Footpath and viewpoints
3. Access field
4. Sports area

Section

1. Freeway
2. Green areas
3. Green corridor
4. Mapocho river

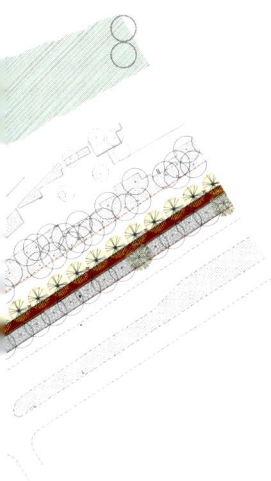

(03) Bike and pedestrian trails, separated by a green belt of river stones and bushes
(04) Sports areas connected by bikeway
(05) Landscape continuity from the Andes to the west

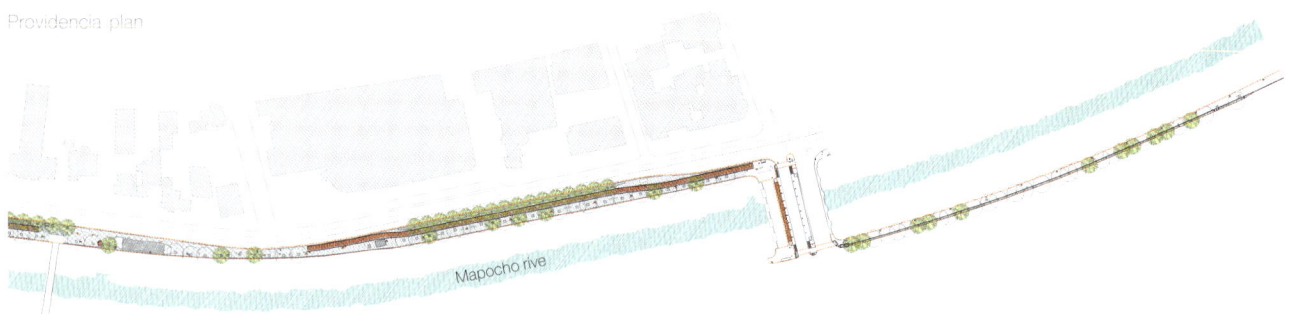
Providencia plan
Mapocho river

06 Promenade along central area
07 Proposed order of proximity to the river of: pedestrians, cyclists and cars
08 Green corridor as seasonal tree-covered backbone

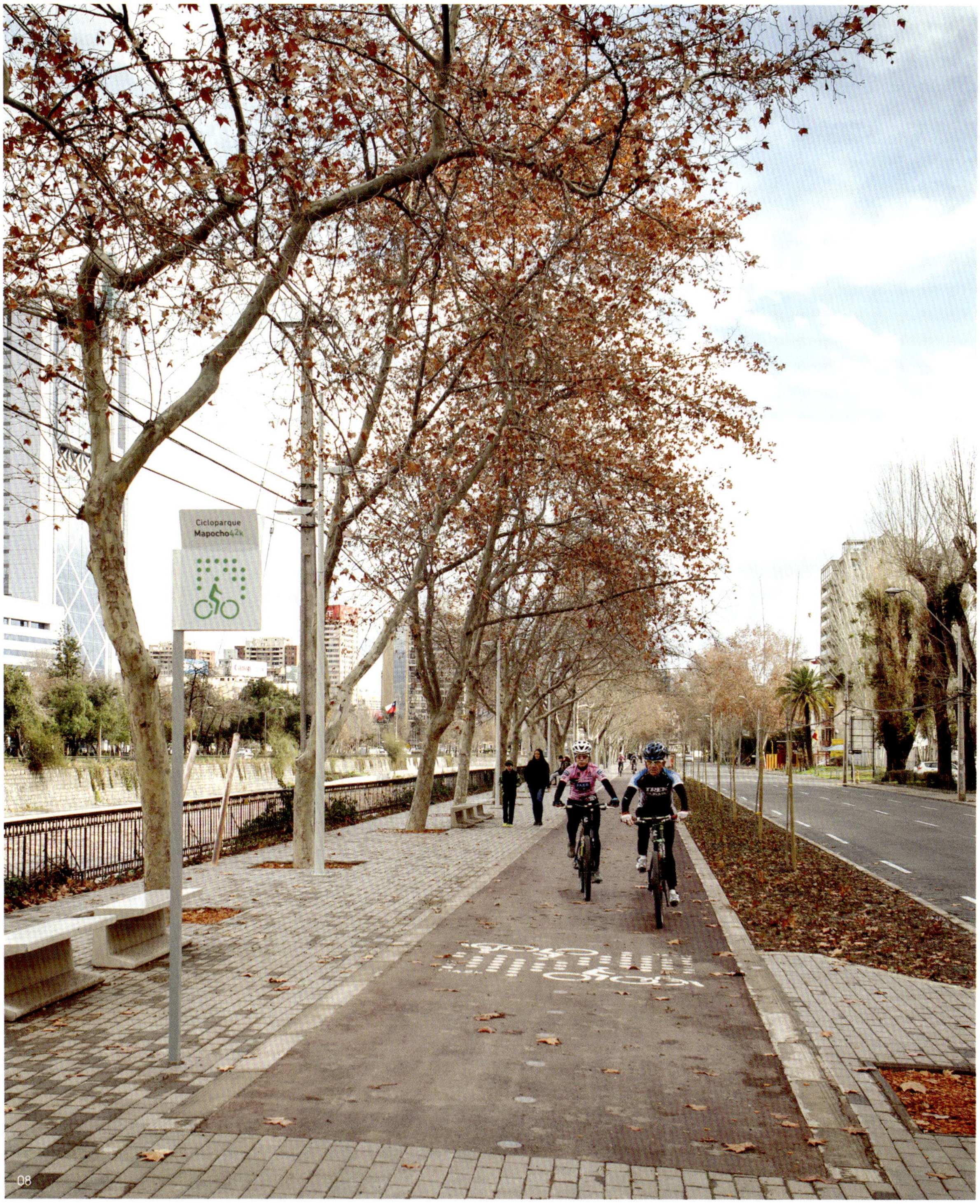

08

1 Park
2 Freeway
3 Mapocho 42K Riparian Promenade

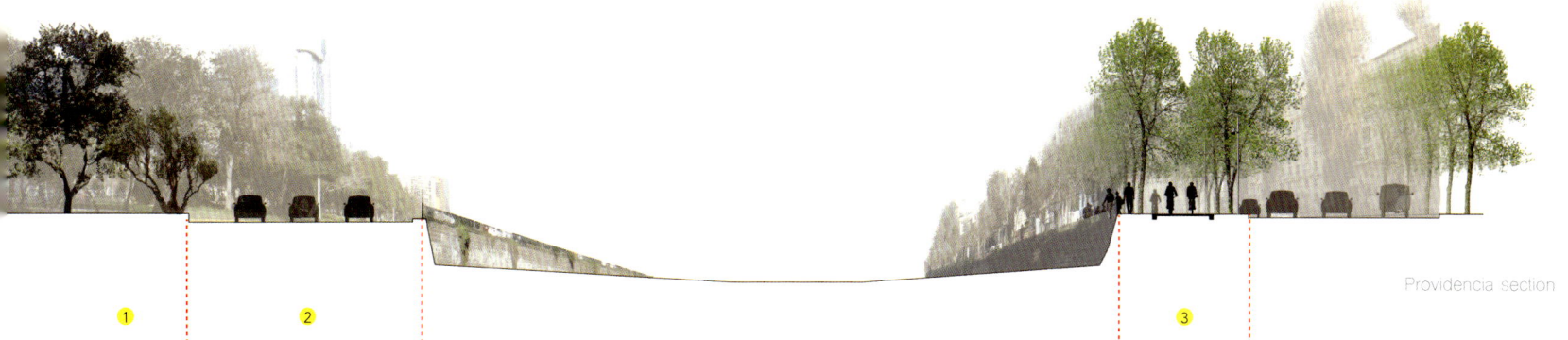

Providencia section

Renewing the Wharf of Austerlitz Port

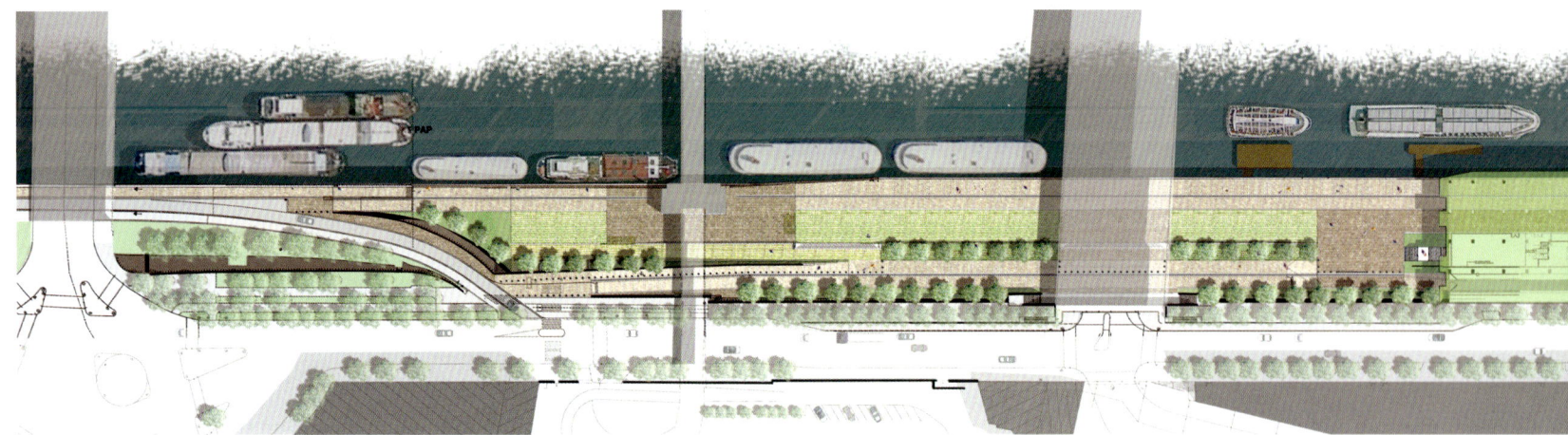

Master plan

● Location / **Paris, France** ● Area / **4.94 acres (2 hectares)** ● Completion / **2012** ● Landscape design / **Urbicus, Jean-Marc Gaulier** ● Photographer /**Charles Delcourt** ● Client / **Ports de Paris** ● Budget / **€4,700,000**

The quaysides of the Seine in Paris, listed as a World Heritage Site by UNESCO, are being reclaimed from the traffic with a strategy that benefits leisure and port activities in the city. The Quai d'Austerlitz is located on the left bank of the Seine, between the Port d'Austerlitz and the Port de Bercy. It is a public area worthy of the city of Paris, constituting one of the sections of the pedestrian city walkway along the banks of the Seine, plus it links the Jardin des Plantes (Botanical Gardens) and the Parc de Bercy. In the 13th Arrondissement, the Quai d'Austerlitz constitutes the façade overlooking the Seine. The aim of the development is to create a new link between the areas bordering the river, offering a place for a walkway right by the water. The Quai d'Austerlitz provides the square in the new, fashionable area in Paris.

This reclaiming of the quayside for pedestrians is achieved by reconciling the constraints imposed by the port activities of the site, particularly in terms of the transportation of passengers and heavy goods, and the presence of houseboats, which means that the area is mostly paved.

The Quai d'Austerlitz is wide with many buildings, and its configuration is not typical. The work on the layout and the choice of materials used in the floors dictate their uses. The sandstone paving, a characteristic ground-covering material used in Parisian open spaces, constitutes the base of the platform. The different finishes mark out and identify the different spaces according to their uses. One large strip, with less foot and vehicle traffic, is set out with turfed areas joining the parts. In summer, this space allows floating restaurants to be installed all along the quayside, creating a dynamic and lively atmosphere. Limestone paving stones line the walkway and trace the outline of the old building demolished for the construction of the Port Charles de Gaulle between the Gare d'Austerlitz and the Gare de Lyon railway stations. The levels have been planned to provide necessary accessibility without interrupting the flow of the Seine when it floods. Large Comblanchien limestone footings mark the changes in level.

This project also involves nature in the city through its relationship with the river and in its use of vegetation in the joining areas. The surfaces with turfed joining areas allow the penetration of rainwater and the development of a measure of biodiversity. A double row of plane trees, a typical sight along the banks of the Seine, frames the traffic lanes. The bottom of the walls lower down on the quayside are planted with Virginia creeper. All of this encourages biodiversity on the banks of the Seine in urban surroundings that are very restricted by human activity. Through this wave of investment, transformation, and development, the Quai d'Austerlitz has become a trendy area popular with the young people of Paris.

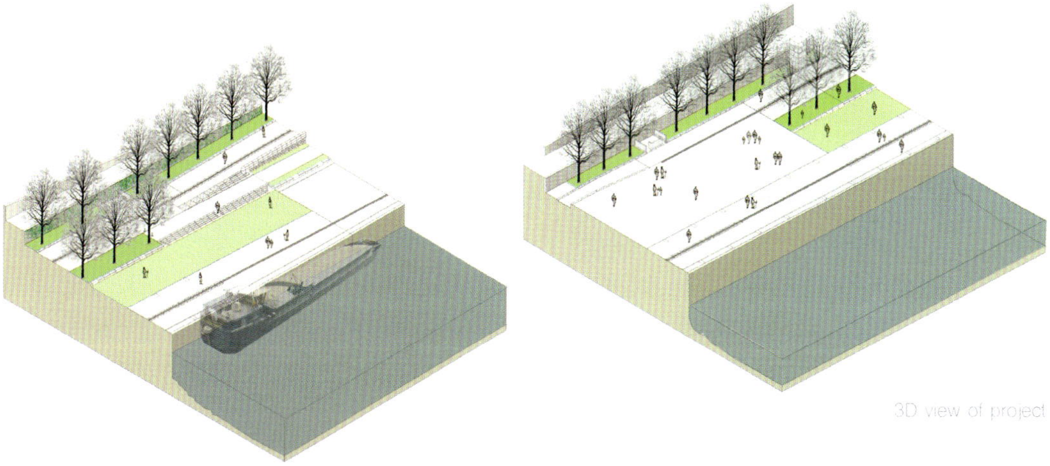

01 Wharf of Austerlitz port renovation as a new link between the areas bordering the river

3D view of project details

02 Shared spaces at the edge of the Seine
03 Nature penetrates the joints of the paving stones
04 The layout of the sandstone paving defines the variety of spaces

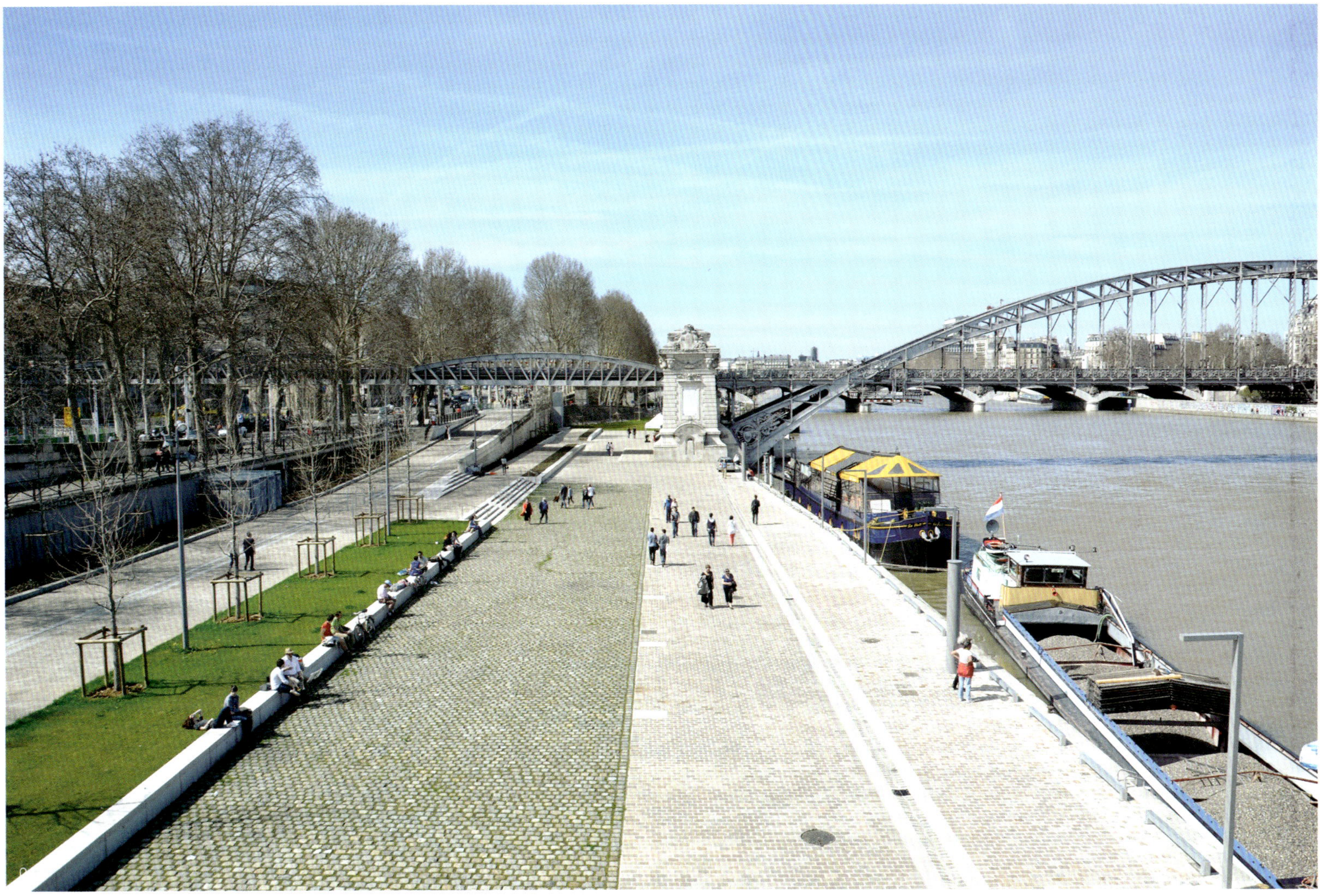

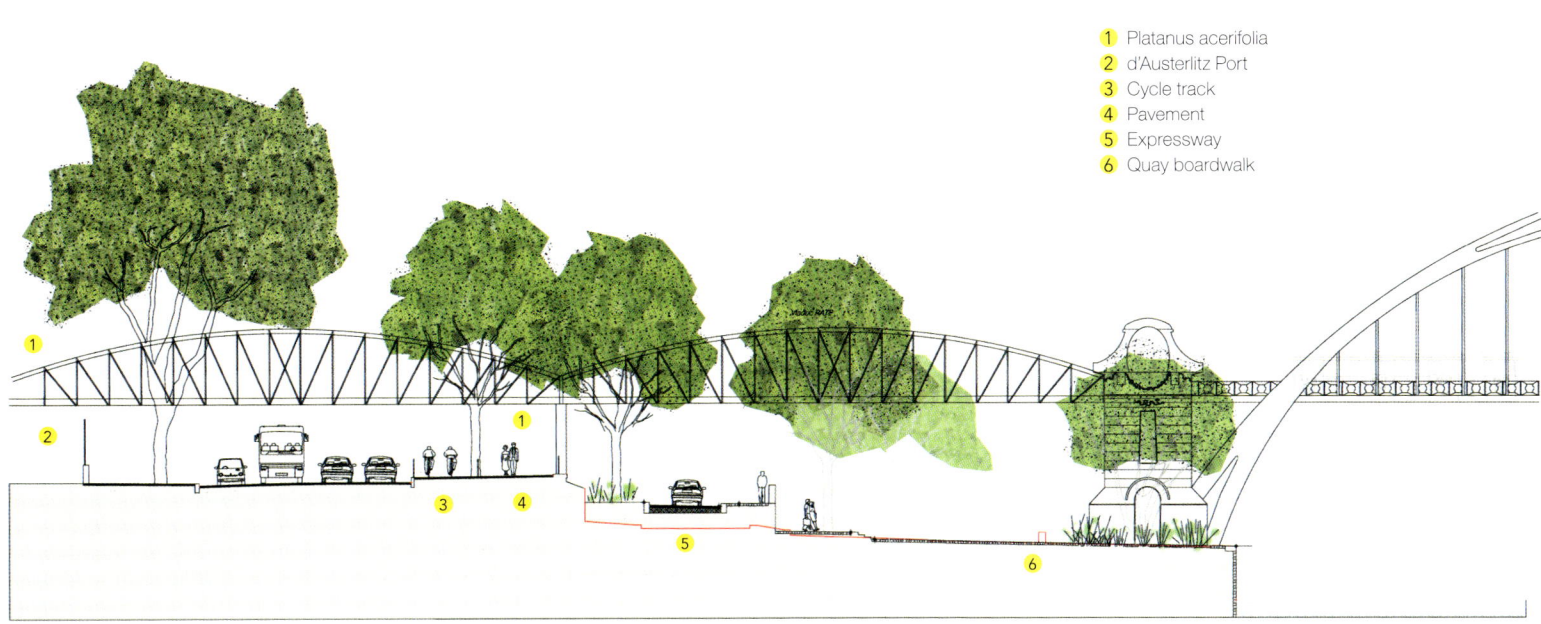

1 Platanus acerifolia
2 d'Austerlitz Port
3 Cycle track
4 Pavement
5 Expressway
6 Quay boardwalk

The project connecting the high quay and the low quay

05 The quay becomes a delightful place to walk
06 The handrails were custom-designed for the project
07 - 09 The barriers are used by the walkers like benches

⑩–⑪ The quay is in the opposite side of the Ministry of Defense
⑫–⑭ The differences in levels are marked by the staircases, facings and benches

A set of terraces toward the Seine

1. Granite edges
2. Sandstone pavements
3. Concrete structure and stone facing
4. Two sandstone pavements
5. Old sandstone pavements

a. Band of plantations
b. Way of transit
c. Pavement
d. Terrace, pedestrian walk
e. Paving strip with grass joints and trees
f. Paving strip with grass joints
g. Service access
h. Pedestrian walk and technical strip

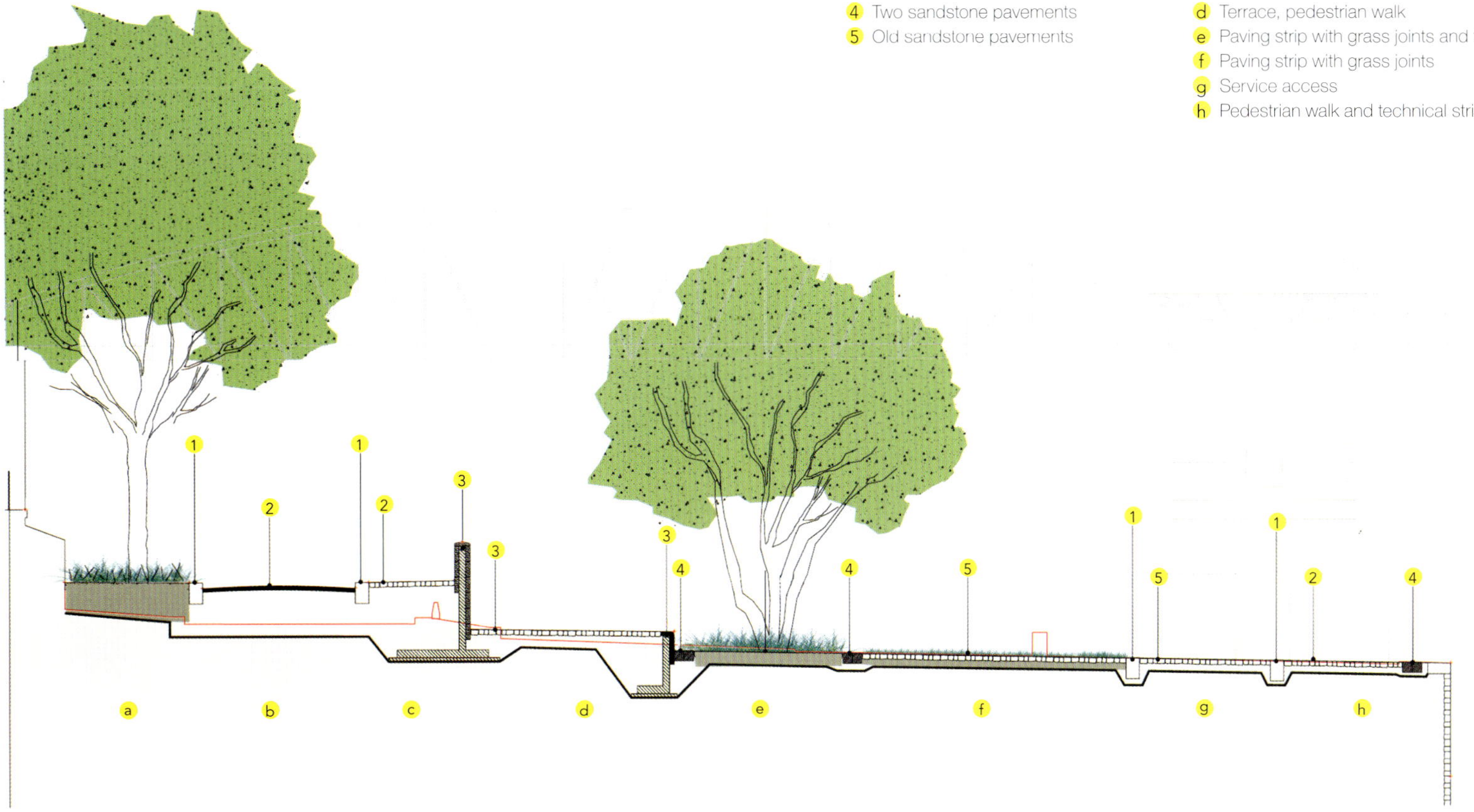

Krymskaya Embankment

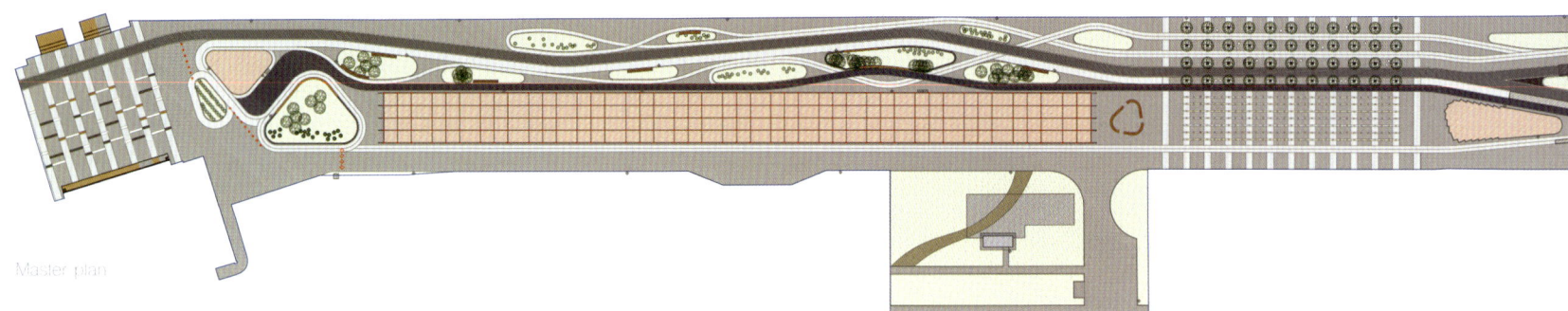

Master plan

- Location / Moscow, Russia ● Area / 11.12 acres (4.5 hectares)
- Completion / 2014 ● Landscape design / Wowhaus Architects
- Photographer / Olga Alekseenko, Yuriy Brazhnikov, Nikolay Vasiliev, Olga Voznesenskaya, Elizaveta Gracheva, Darya Osmanova, Alexander Minchenko, Darine De'Reviere ● Client / Muzeon, MosGorPark, Moscow Department of Culture ● Budget / RUB2,700,000,000

Krymskaya Embankment is a project re-establishing one of Moscow's central embankments as a public space by turning it from a disused highway into a public recreational transit zone for pedestrians. Despite its central location, the embankment has been avoided by residents as it was an unengaged road with a petrol station, which was used as a dumping ground for dirty snow. After its transformation, the Krymskaya Embankment became the first year-round landscape park in the center of Moscow. The wave-shaped multilevel layout can be used for walking, cycling, or roller-skating in the summer, while in winter it is a perfect setup for sledding, skating, or skiing.

The problem identified by Wowhaus is that the Krymskaya Embankment is in fact the missing link in the long pedestrian route around the city center. A study showed that making the road a pedestrian-only zone would only marginally increase traffic loads on the nearby streets. By making the embankment a pedestrian zone, Wowhaus proposed to join a string of parks with the city's historical center. Thus, the city acquires a 6.21-mile (10-kilometer) stretch of walks and cycle routes through succeeding parks forming a so-called "Green Loop."

From the southeast, the embankment borders Muzeon Park—a traditional place for solitary walks near the modern building. So the challenge was to create a park with a transit capacity to accommodate visitors of nearby parks (pedestrians, skaters, and cyclists), which at the same time honored the traditional tranquility of the nearby Muzeon Park. The designer responded to the challenge by designing a green recreational transit zone.

The linear structure of the park paths along the waterfront is supplemented with a series of hills and waves closer to Muzeon Park. This allows for seamless redistribution of visitors from Muzeon Park and regular passers-by. The central design element of the embankment is the wave: wave-shaped benches together with wave-shaped walking and cycling paths and facilities create an artificial landscape. The park zone is divided into four parts: an area in front of the bridge, an artists' zone around a "Vernissage" pavilion, the Fountain Square, and "Green Hills." In planning each zone, the view from the other bank was also considered.

The bridge space is a transit zone connecting Gorky Park with the Krymskaya Embankment. It has become a popular spot and also provides shelter from the rain now that a stage and two wooden amphitheaters have been built. Twenty-eight artificial rock and metal benches illuminated from the inside are scattered along the way as an amenity for pedestrians and cyclists from Muzeon to Gorky Park.

(01) Wave-shaped artist pavilions

01

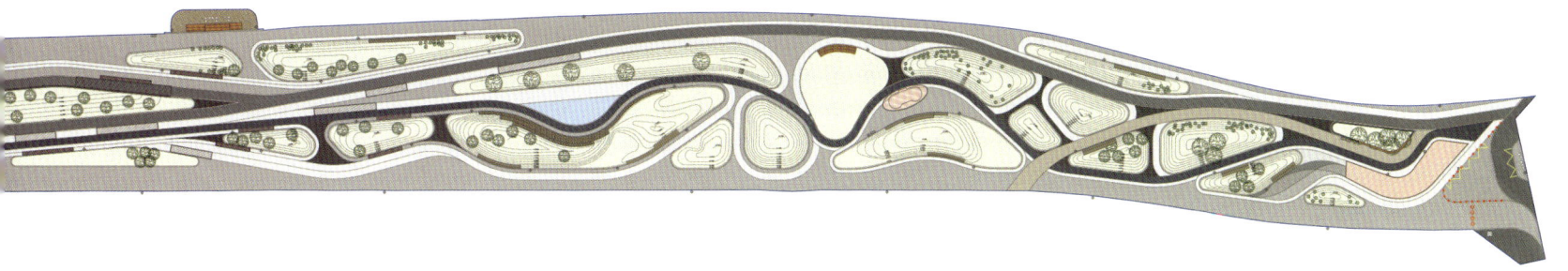

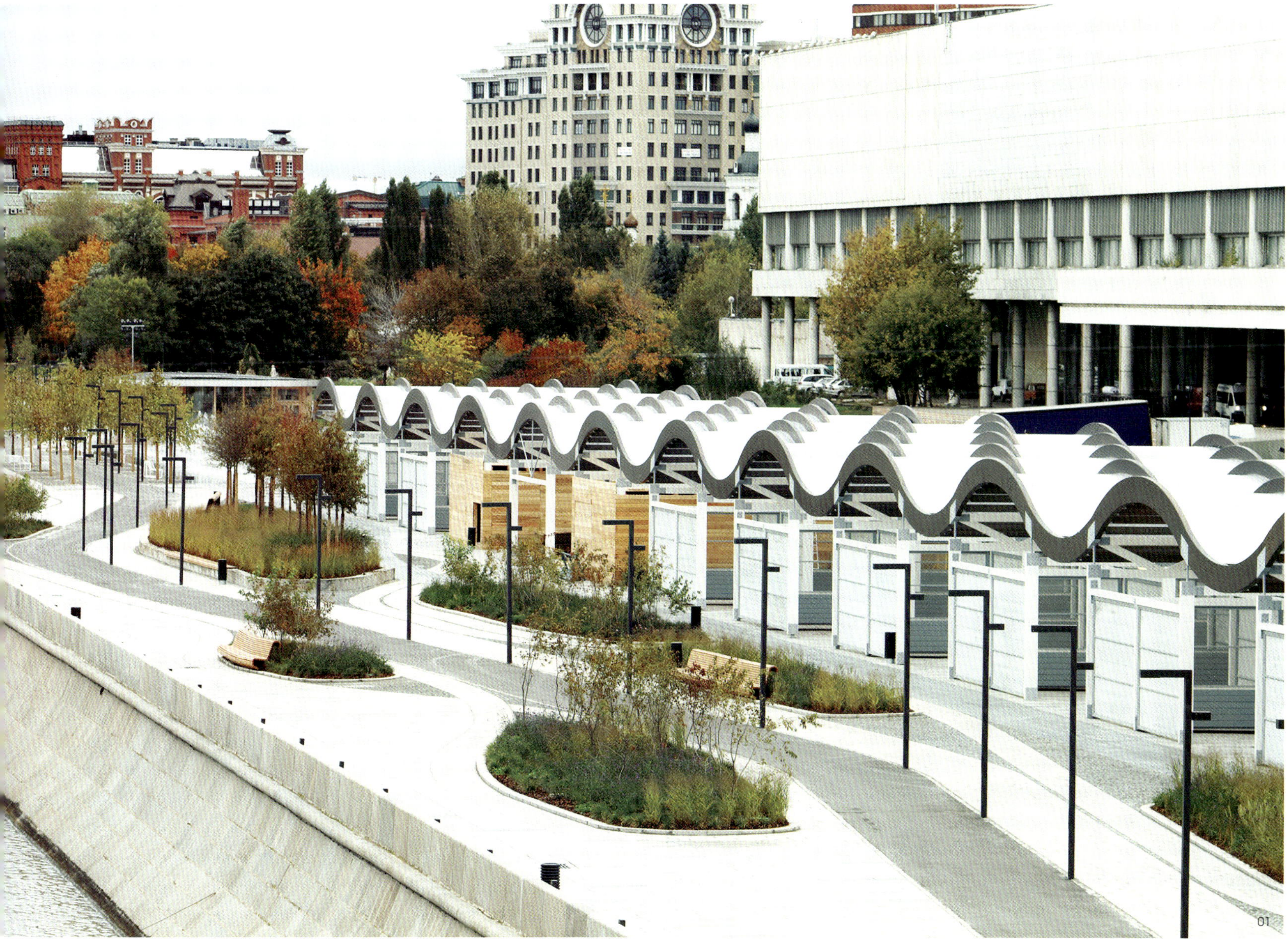

Waves details of cycling and walking route

There is a special pavilion for art fairs (by Asse architects) that have replaced a chaotic exhibition area.

The fountain zone is the central element of the new park that is separated from the river by a linden alley. A fountain jet, 196.85 feet (60 meters) long and 45.93 feet (14 meters) wide, is one of the options of the so-called "dry" fountains when the edge of the water is level with the paving. It creates an opportunity to use it as a square for city events.

When planning this part of the pedestrian route, special attention was paid to the artificial landscape and plantation. Hills designed for walking and resting were furnished mainly with steppe plants. Trees and bushes with decorative crowns such as lindens, hawthorns, rowan trees, and ornamental apple trees were planted on hills where one can contemplate and admire the scenery. The artificial relief is accentuated by wave-shaped wooden benches and beach beds that are "cut" into the hills between walking lanes.

To make the park accessible and attractive for guests twenty-four hours a day, planning takes into account night-time illumination, especially the point lighting of certain landscape elements. Ornamental lamps that are installed in groups among plants on the hills illuminate the area and create a striking visual effect. All the lanes are illuminated as well so that pedestrians and cyclists do not get lost.

(02) Pavilion on the embankment is also part of the walking and cycling route
(03) Wave-shaped elements of the cycling and walking route
(04) Krymskaya Embankment during the daytime

05 Benches
06 Walking and cycling routes
07 Fountain zone
08 Pedestrian route and green hills
09 Transit zone between Krymskaya Embankment and Gorky Park

Westhaven Promenade

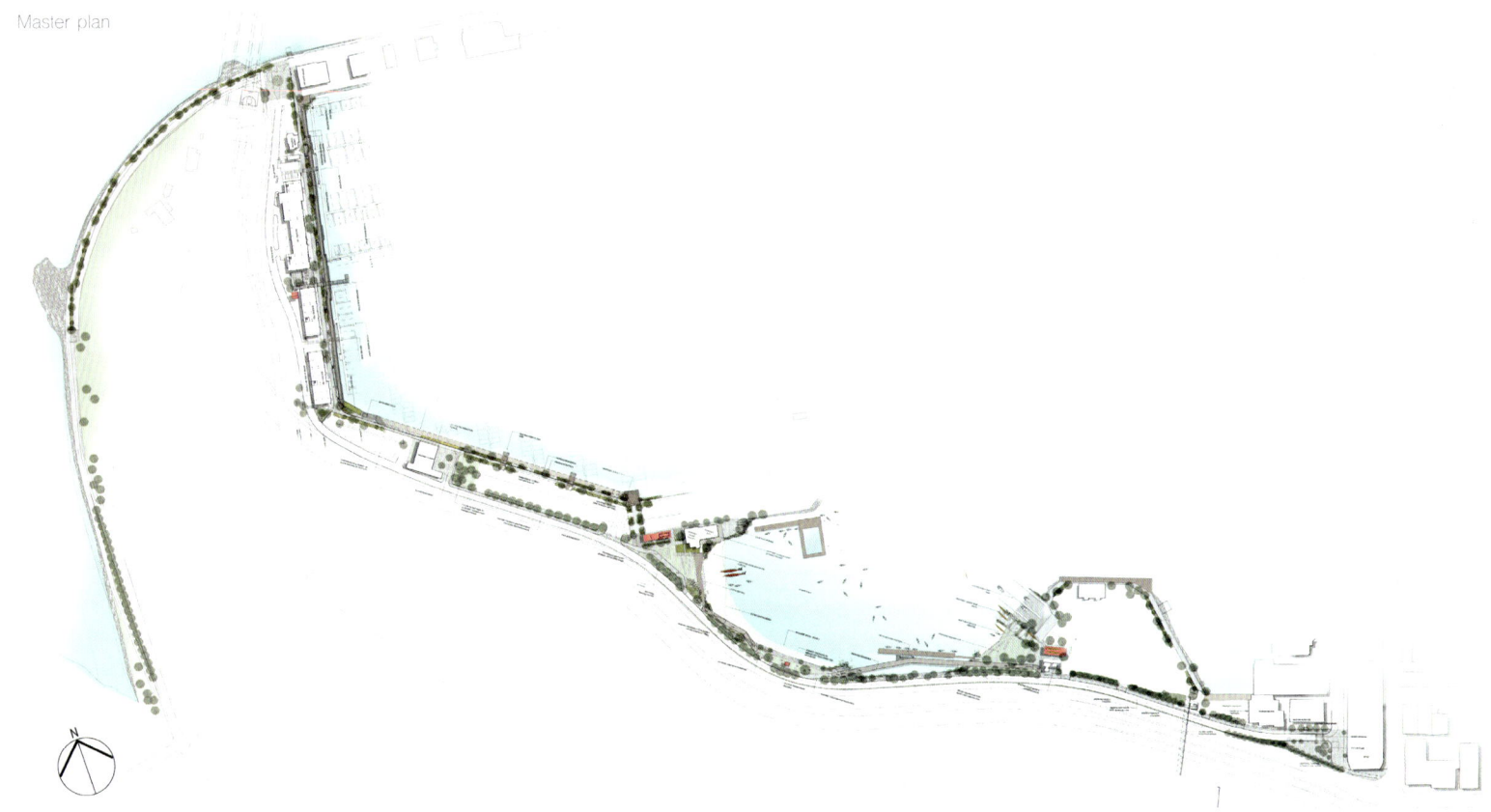

Master plan

- Location / **Auckland, New Zealand** ● Length / **0.93 miles (1.5 kilometers)** ● Completion / **2015** ● Landscape design / **ASPECT Studios, Architectus and LandLAB in collaboration** ● Photographer / **Jonny Davis** ● Client / **Waterfront Auckland** ● Budget / **NZ$7,000,000 (1st stage)**

Westhaven Promenade is a publicly accessible shared bicycle and pedestrian waterfront promenade. Located in Westhaven, Auckland, the 0.93-mile (1.5-kilometer) promenade links Westhaven Marina to the city center.

The new promenade is a significant contribution to Auckland's public realm. It provides a continuous and safe pedestrian and cyclist link between the city and Westhaven Marina—a formerly disconnected area—and improves the amenities and recreational opportunities of the city-center waterfront. A network of elevated decks, new parks, and water interfaces are set among the interpretive elements that reveal the local narratives of place and occupation.

Westhaven Promenade celebrates the harbor rhythm of headlands and bays, passing through distinctly different zones and offering moments of unique experience. Vantage points and places to congregate capitalize on significant views and vistas, and offer a range of pedestrian experiences along the promenade. The project also informs a broader strategic move of adding to a continuous 24.85-mile (40-kilometer) "Waitemata Harbor Loop" that will have both local significance and international appeal when completed.

The new promenade restores access to the public harbor through improving safety of the waterfront and providing a continuous and accessible public link between unique destinations, parks, and beaches. The promenade increases the possibilities for recreational use of the harbor. Beaches are reinstated, providing key destinations and places for people to gather on the city's waterfront. The promenade links to the revitalized "Marina Village," where new commercial and retail activities provide a public and lively destination adjacent to the port infrastructure.

① Marina promenade south view
② Pedestrians strolling down the south side of the marina promenade

03

04

03 Marina promenade view north
04 Marina promenade detail view
05 Promenade over bay
06 Promenade and seating over bay

07 Marina promenade north edge
08 Viewing platform
09 St Mary's Bay promenade and landscape works
10 New engraved concrete seating steps at St Mary's Bay
11 Seating steps

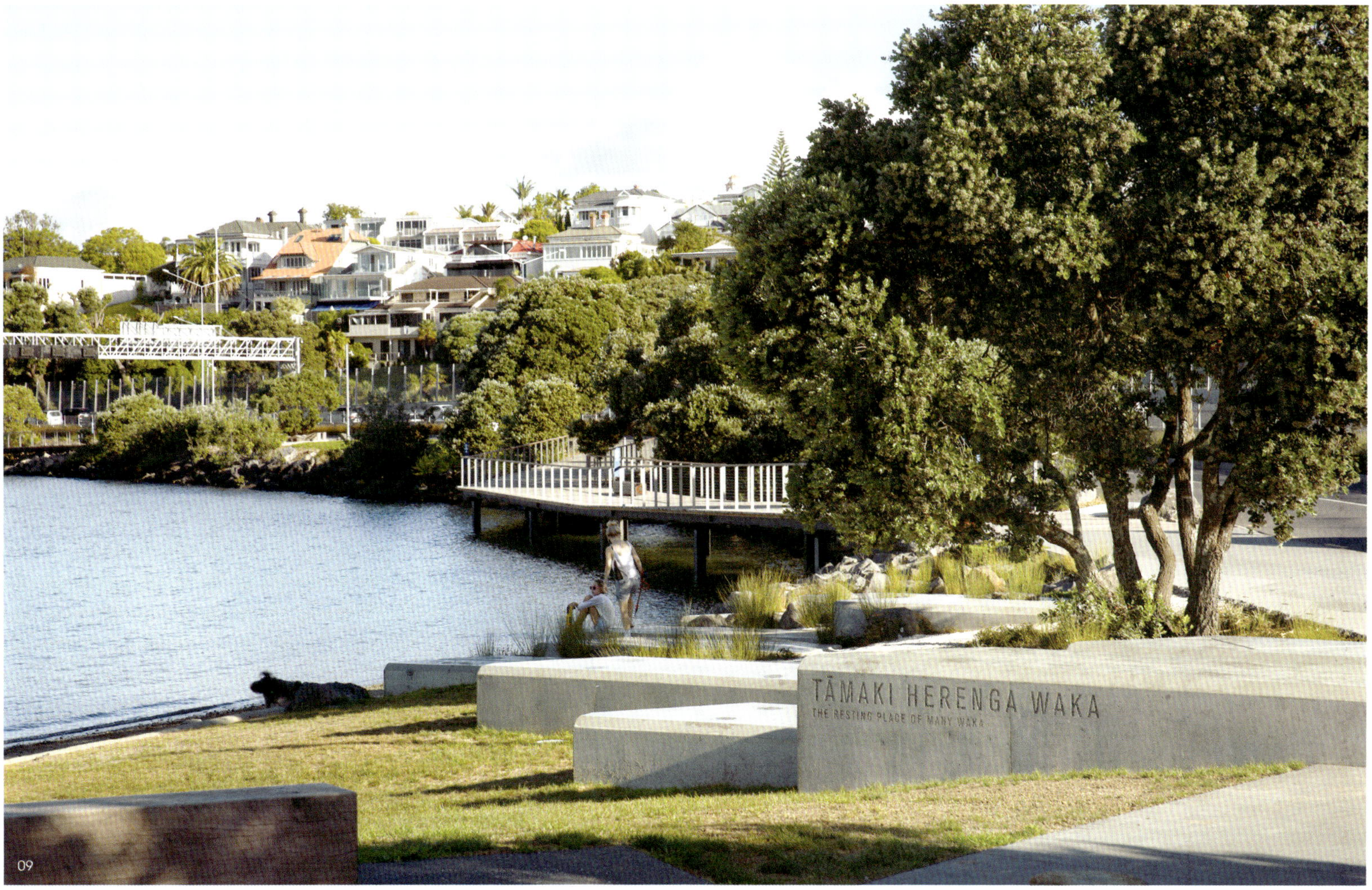
TĀMAKI HERENGA WAKA
THE RESTING PLACE OF MANY WAKA

09

TĀMAKI HERENGA WAKA
KOTAKEREHEA
THE SPLIT CANOE HULL

10

11

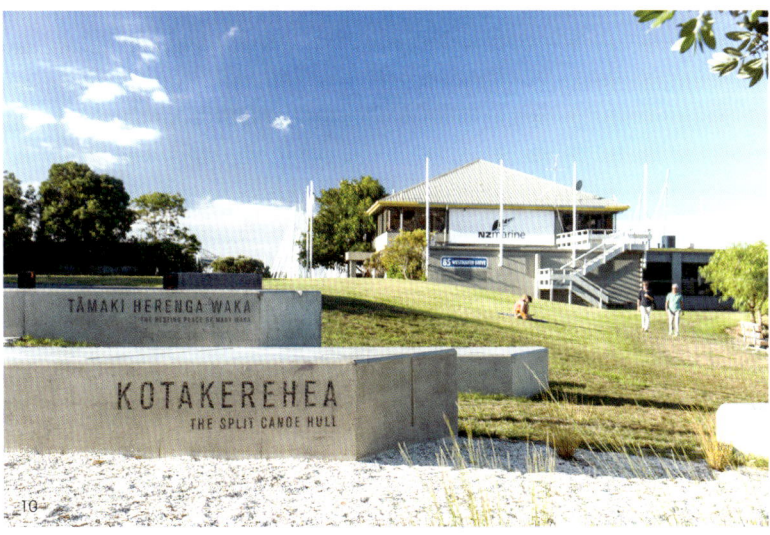

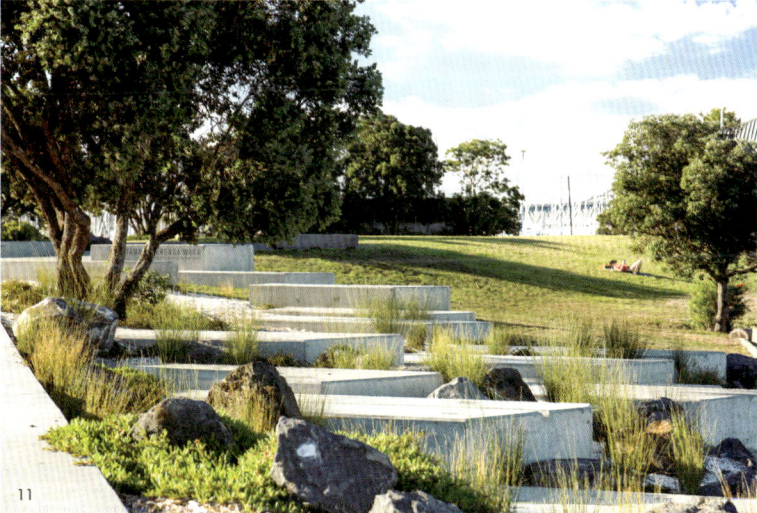

La Perouse Headland Coastal Walk and Loop Road

Master plan

- Road and parking
- Shared zone
- Coastal walk
- Footpath
- Landscape zone
- Separation strip and bollards
- Road reserve boundary

● Location / Sydney, Australia ● Area / 25.69 acres (10.4 hectares) ● Completion / 2012 ● Landscape design / Corkery Consulting Pty. Ltd. ● Photographer / Corkery Consulting, Denliz ● Client / Randwick City Council ● Budget / AU$2,500,000

Located on the northern shores of Botany Bay, La Perouse Headland provides spectacular cliff-top views extending out to the Pacific Ocean. The 25.69-acre (10.4-hectare) headland contains many listed cultural heritage points of interest that are protected through the national park. Heavy use and uncontrolled parking along the loop road had significantly affected the quality of La Perouse Headland. Randwick City Council therefore decided to implement a major program of upgrading, including a new section of coastal walkway that provides safe and comfortable access for pedestrians around the headland, and connects to a regional coastal path system that will ultimately extend from Bondi Beach to La Perouse and link to the proposed Botany Bay Trail as part of the Sydney Metropolitan Regional Recreation Trails Framework.

The design enhances the setting of heritage structures and sites on La Perouse Headland while not detracting from them. Views from the headland have been retained through minimal use of vertical structures. Meanwhile, it was necessary to replace the old steel crash barrier with attractive, precast concrete bollards that are strong enough to protect vehicles from going over the cliff, while allowing pedestrians to gain access from their parked cars to the new section of coastal walkway. Particular attention was paid to pedestrian movement within the adjoining areas of the national park and the coastal walk.

Visual analysis

1. View between headlands to open sea
2. Water view partly blocked by parked cars
3. View dominated by Bare Island
4. Open view across bay to Kurnell
5. Water view partly blocked by parked cars
6. View across bay with grass open space in foreground
7. View of Yarra Bay with Port Botany beyond
8. Confined view across Yarra Bay
9. Visually confined by building along east edge and slopes to west
10. Visually confined by embankment to east
11. Intersection with views to north, west and south
12. Visually enclosed by vegetation to the south and by parked buses and cars to the north

(01) Precast concrete bollards and seating adjoining loop road and parking

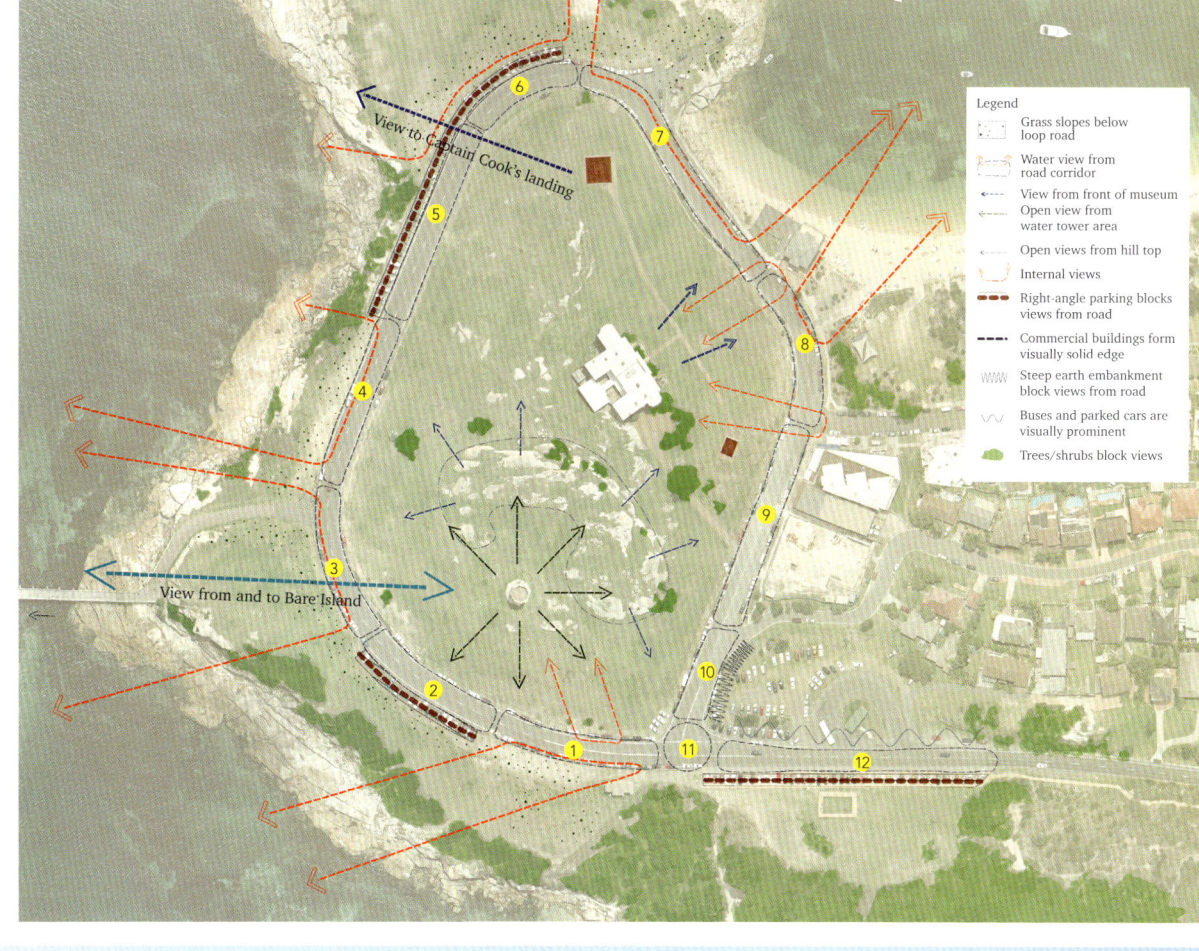

Construction of the section of coastal walk along the edge of the road corridor was carried out with minimal site disturbance. The original dirt track, rusting steel crash barrier and uncontrolled roadside parking were replaced with a 2296.58-foot-long (700-meter-long) and 9.84-foot-wide (3-meter-wide) colored concrete path, precast bollards, seats, a one-way section of road with organized parking for 220 vehicles, precast wheel stops and edging as well as plant beds. The new parking layout retains the total number of parking spaces while creating a more efficient and safer environment for pedestrians and cyclists.

Creation of a one-way section of road with roundabouts at each end allows for safer traffic flows and space for a separate pedestrian coastal walkway. The use of 245 concrete bollards not only provides physical separation between vehicles using the loop road and pedestrians using the coastal walk, but also provides extensive seating and avoids use of standard curb and gutter. They also form a distinctive urban element and maintain open coastal views.

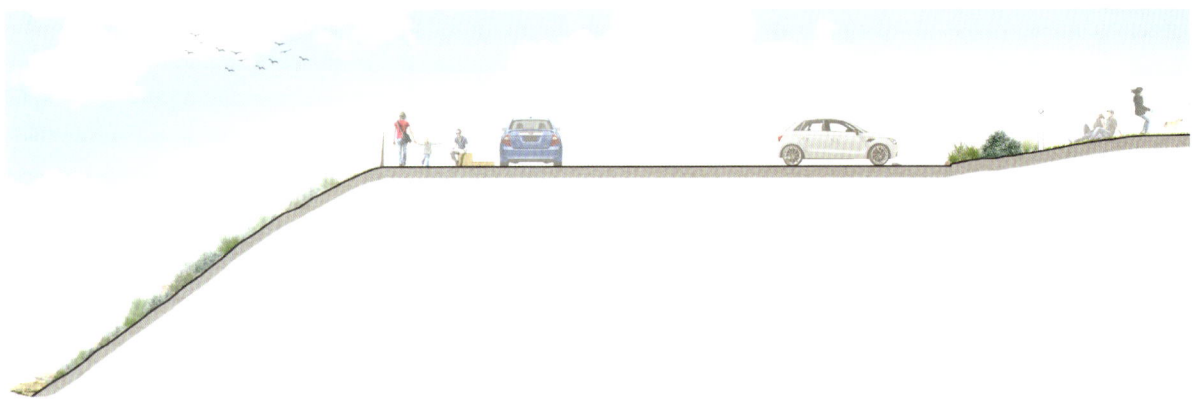

Typical cross-section

02 Rain garden with native plants along the coastal walk
03 Precast concrete wheel stops and bollards along the coastal walk
04 Precast concrete bollards separating parking from the coastal walk

Given the very high cultural heritage importance of the La Perouse Headland, considerable research was carried out into the history and significance of the site to ensure the design respects these values. This project achieves outcomes in both environmental and social sustainability. The use of species indigenous to the adjoining section of the national park, with passive irrigation provided by runoff from the adjoining road surface, not only reuses stormwater but also produces water-quality improvements and contributes to biodiversity of the headland and adjoining sections of Botany Bay.

The significance of the La Perouse Headland upgrade project is reinforced by current research, which is providing growing evidence that high-quality public domains make a major contribution to community health and sense of well-being. The coastal walk is a major new pedestrian facility that encourages walking. Upgrading 2624.67 feet (800 meters) of road surface of the Loop Road allows an increasing number of cyclists to safely enjoy movement through the area.

Picnicking, fishing, scuba diving, and visiting the museum can now be enjoyed, and outdoor dining is an option at the two new cafés along Endeavor Avenue at the northern edge of the Loop Road.

05 Precast concrete bollards and wheel stops adjoining the coastal walk
06 Memorial to Captain La Perouse with bollards at pedestrian crossing
07 Sandstone curb and gutter with exposed aggregate strip adjoining car parking
08 Coastal walk, precast concrete bollards and seatings

Typical plan and cross-section

1 Coastal edge
2 Safety rail
3 Coastal walk
4 Separation blocks
5 Parallel parking
6 Road
7 Right-angle parking
8 Curb
9 Drainage line
10 Sandstone boulders
11 Aris rail
12 Existing slope
13 One-way traffic

09 Loop road with museum and La Perouse Memorial as part of the skyline
10 View from coastal walk to heritage watch tower
11 Cyclists entering loop road at La Perouse Headland
12 Native species planting along coastal walk

Kalvebod Waves

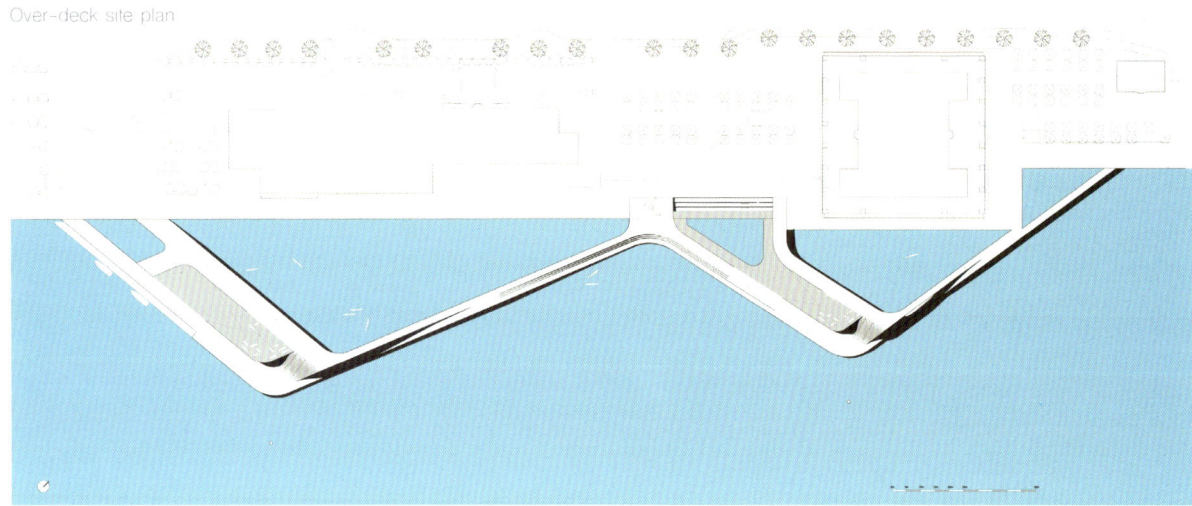

Over-deck site plan

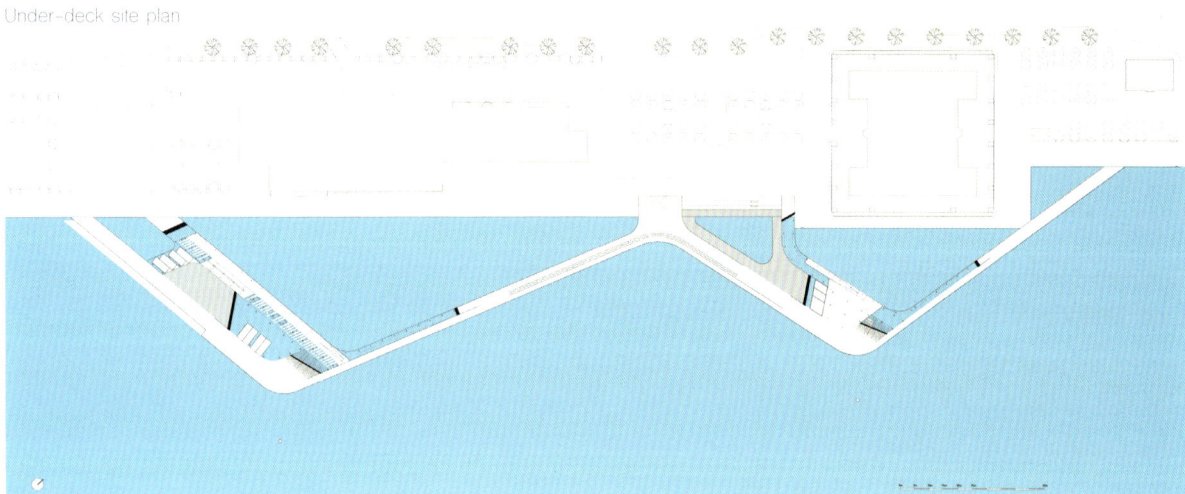

Under-deck site plan

● Location / **Copenhagen, Denmark** ● Area / 1.04 acres (0.42 hectares) ● Completion / **2013** ● Architect / **Urban Agency, JDS** ● Photographers/ Urban Agency, Pernille Enoch, Ursula Bach ● Client / **Copenhagen Municipality** ● Budget / €7,000,000

With the opening of the Kalvebod Waves at the harborfront of Copenhagen, a central part of the inner harbor has finally become accessible and attractive to the public. This new public space on the water gives the harbor a new dimension as a promenade in the center of the city. Historically, this part of the harbor was devoted to industrial activities. Then, in the 1980s and 1990s, the city sold the land, and the area was developed into a harsh and mono-programmed harborfront, leaving its quay barren and open to strong winds, devoid of any public life.

When addressing this infamously gloomy and desolated side of the harbor, the architects focused on two major design priorities: to create urban continuity and to locate the new structure on the sunny spots by the water.

Facing southeast, the quay has fantastic sun conditions until the early afternoon, but to attract the public after working hours, the architects needed to study the course of the shadows cast by the huge buildings along the promenade. The two new main piers are located in the shadow-free zones and designed as activity hubs on the water. The project then reconnects those islands to the urban network, where they blend naturally into the city's infrastructure. Two bridges link the tip of the main piers back to the old quay, creating two inner water basins for water-sport activities.

01 Bird's-eye view

Just like the plan, the project is making waves in the third dimension: On the pier, visitors can explore the waterfront from different levels and enjoy its amazing views. People can get close to the water and stick their feet in or have a swim, or they can get 16.40 feet (5 meters) up, as if they were sailing on a boat. The new development invites people to take a walk on the promenade to enjoy an exciting and active public space.

The design allows for many different activities to take place. The architects' intention was to create a frame for an unknown content, a place for public life to unfold, and the unexpected to happen. It proved to be the right strategy. People adopt the Kalvebod Waves in their own ways, beyond what anyone could have imagined in advance.

In terms of its materiality, the project is inspired by industrial harbors. Robust materials such as planks from local oak trees, broomed concrete, and rusted steel are known as harbor materials. The light and elegant stainless-steel guard-railing integrates the lighting, and the design echoes the detailing of sailing boats. The project is designed and built according to the "cradle to cradle" building philosophy. The used materials are defined in "cradle to cradle" terms of chemical contents, effects on air, soil and water, and effects on human health from manufacturing through use and recovery. The project has been designed and built with the help of Building Information Modeling (BIM), which has improved the quality of the building systems, products, and processes as well as the total material consumption. All components are recyclable, and the project can be disassembled and reused to a large extent.

02 Public space on the water
03 View of the waves in vertical dimension
04 View of the waves at horizontal level

05 A new waterfront for cycling and strolling
06 Recreation activities
07 Wooden deck
08 People relaxing on the wave
09 View from the water

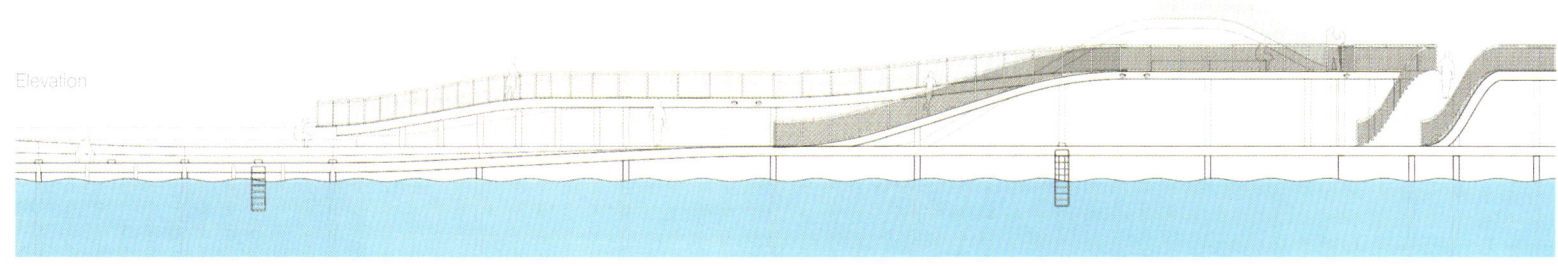

Elevation

⑩–⑫ View of detailed wave elements
⑬ People exploring the harbor on different levels
⑭ The waves provide a safe and exciting cycling experience on the waterfront

Central Dandenong Lonsdale Street redesign and upgrade

Master plan

- Existing buildings
- Development sites
- Greenspace
- New and existing trees
- Pedestrian priority
- New development
- Key buildings
- New roads
- Metrovillage

● Location / **Melbourne, Australia** ● Area / **17.3 acres (7 hectares)** ● Completion / **2011** ● Landscape design / **BKK Architects, Taylor Cullity Lethlean Landscape and Urban Design** ● Photographer / **John Gollings** ● Client / **Places Victoria** ● Budget / **AUS$200,000,000**

Lonsdale Street in central Dandenong is the first key infrastructure project delivered as part of the state government's Revitalizing Central Dandenong (RCD) Initiative. The RCD Initiative seeks to restore central Dandenong as the capital of Melbourne's southeast and bring new energy, activity, and amenities to the heart of this richly diverse urban center.

Central Dandenong has a unique cultural richness, a dynamic produce market, performing arts precincts, and distinctive retail sectors, yet the economic decline of the city over many years took its toll on the overall civic character and public realm experience. Lonsdale Street was historically a prosperous retail spine, but in recent years has developed into a major arterial route dissecting the retail heart and creating a significant physical and psychological barrier to the city. BKK/TCL's approach to urban design projects of this magnitude is curatorial, recognizing that successful urban design should not be concerned with a fixed plan but should instead offer key ideas that are fundamental catalysts for change. Equally, this project furthers BKK/TCL's ongoing investigations into place-making, and strategies that build upon local character, offering positive solutions for change to strengthen and empower communities.

Lonsdale Street was informed by an extensive consultation undertaken by Places Victoria (formally Vic Urban) and the City of Greater Dandenong, which identified valued physical and cultural characteristics of the city and provided the aspirational components of the brief. This study was supplemented by exhaustive mappings and master planning for the city to ensure the urban design for Lonsdale Street was transformational, yet built upon the distinctive qualities of the setting.

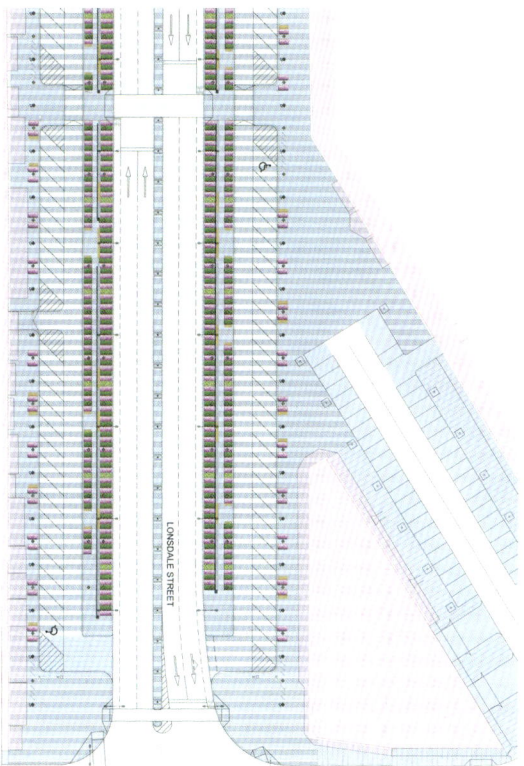
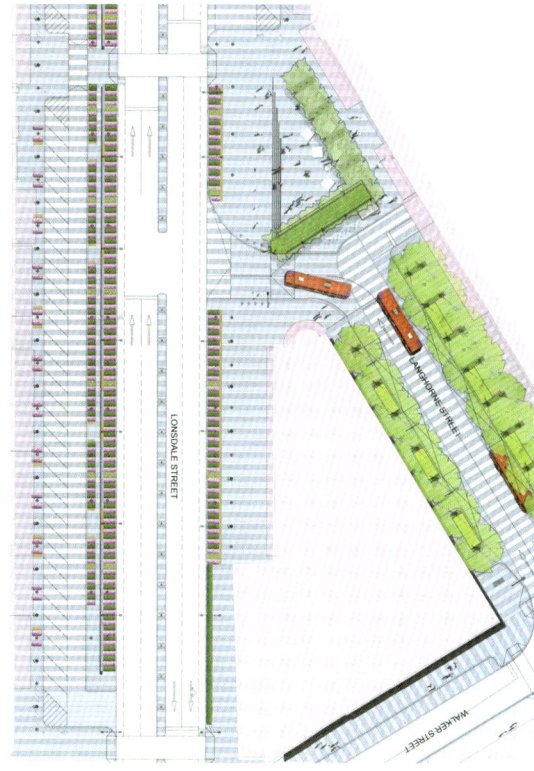

Lonsdale Street plan

01 View to shared pedestrian zones
02 Aerial view

The vision for Lonsdale Street was built upon a number of strategic moves:
- Connections: Instead of being a significant barrier, Lonsdale Street became a key connecting thoroughfare, fostering clear and legible street connections to each of the city's key public assets.
- Street life: Efforts were made to create a memorable, activity-filled boulevard, with retail options, pocket parks, and civic plazas.
- Knitting into the urban morphology: Lonsdale Street was structured to be built upon the distinctive urban structure of the city, reinforcing existing patterns.
- Protecting valued urban places: Lonsdale Street and its adjacent precincts were identified as a significant opportunity to curate the ongoing retention of cultural destinations, and create new opportunities for urban places and activities to develop.
- Investment and design excellence: Opportunities were opened for investment and further development via the creation of a rich and enduring public realm experience.

Lonsdale Street was conceived as a grand boulevard with a pedestrian focus. Through traffic is concentrated into a central band defined by four rows of trees. Adjacent to the retail frontages, broad tree-lined plazas, shared traffic zones and linear gardens provide a pedestrian realm of generosity and distinction. The design is an example of an interdisciplinary approach to the construction of the city involving expertise across a wide range of disciplines and continuous liaison with key stakeholders.

03 Stripped pavement unifies the street
04 Lighting feature along the entire length of the street represents the ethnic diversity of the local population

Axonometric

Section

Plan

03

04

05 Detail of "outdoor rooms"
06 The landscape provides for many varied experiences
07 The civic square provides a space for activity and rest
08 Detail of the square

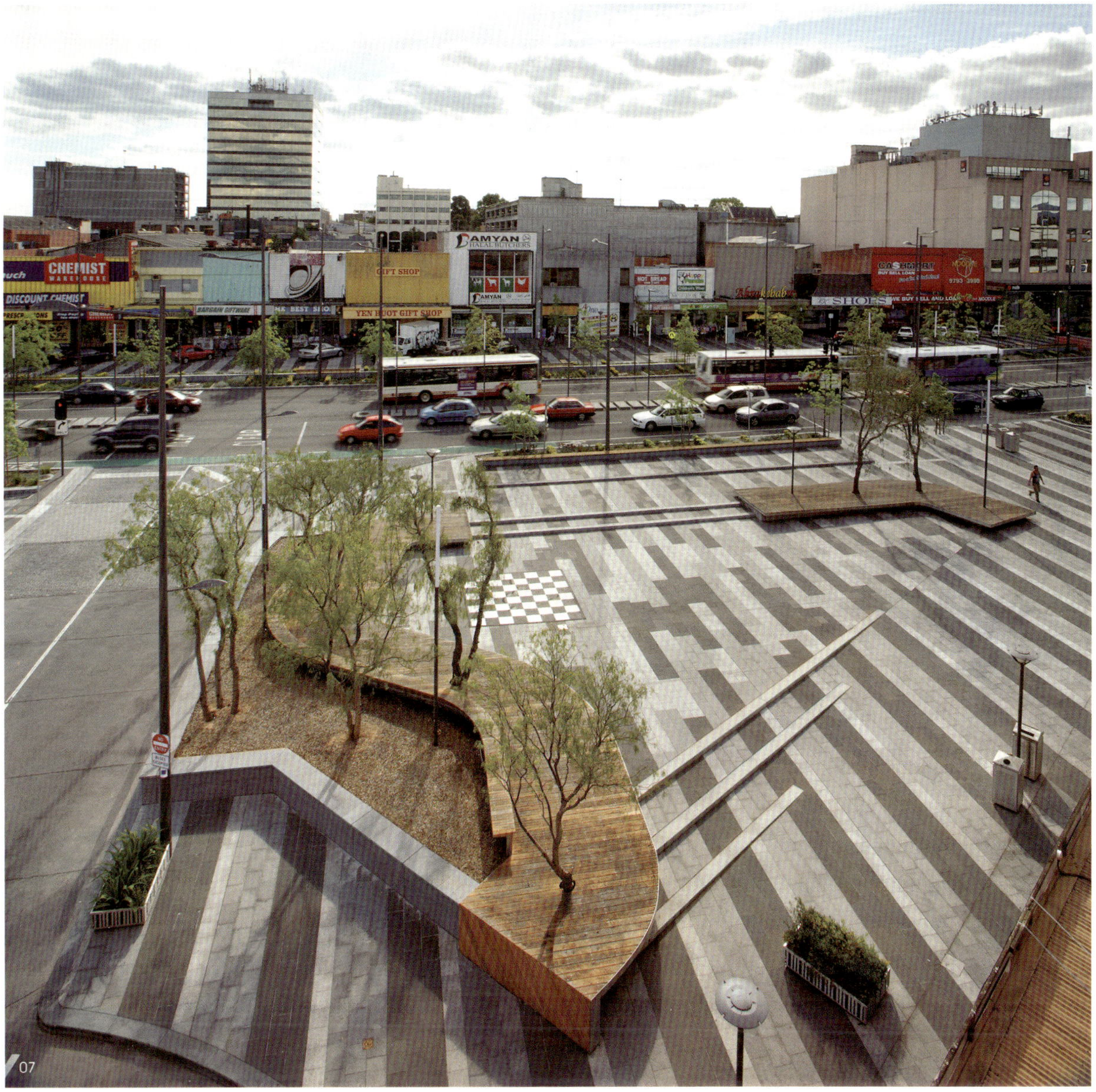
CHEMIST
DISCOUNT CHEMIST
GIFT SHOP
YEN HOOT GIFT SHOP
DAMYAN
HALAL BUTCHERS
07

DAMYAN
Happy Panda
08

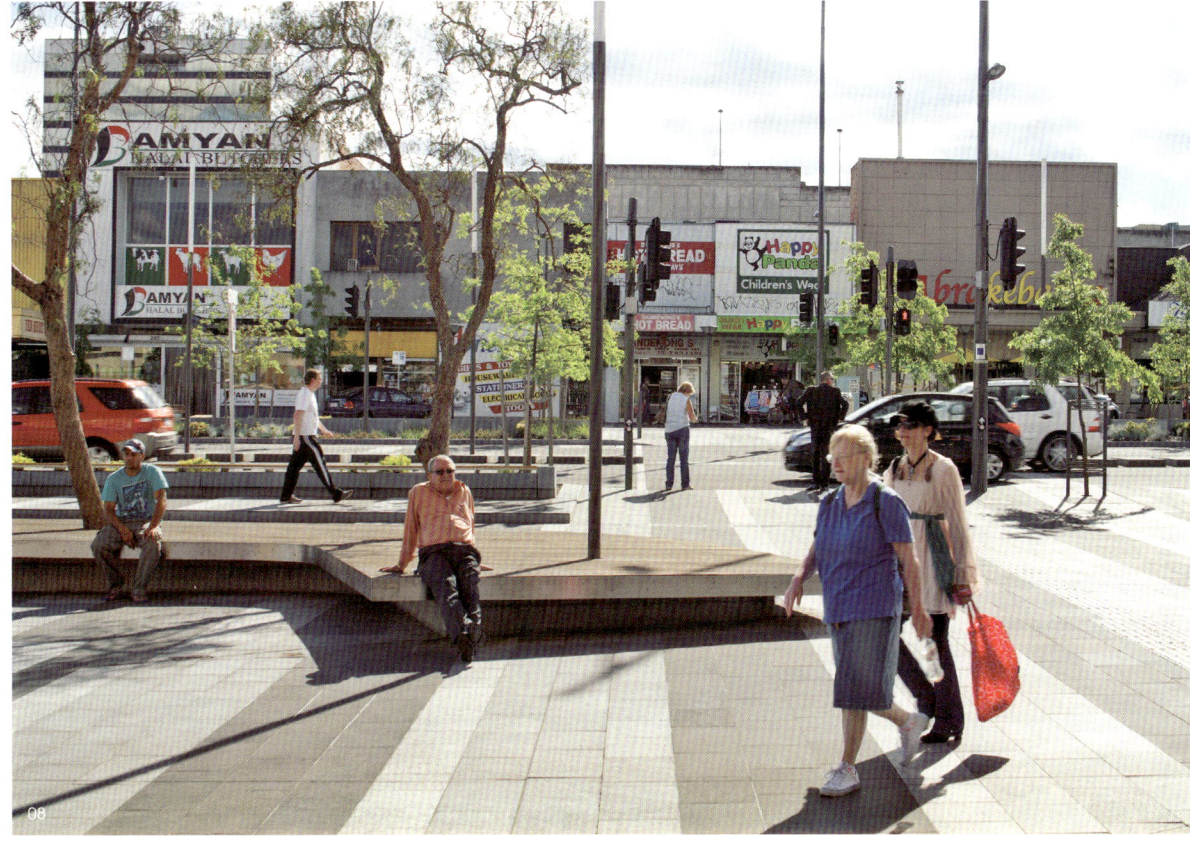

Padua Train Station Square

- Location / Padua, Italy ● Area / 7.29 acres (2.95 hectares)
- Completion / 2013 ● Landscape design / CZstudio associati
- Photographer / CZstudio associati, Guido Ranieri Da Re ● Client / Municipality of Padua ● Budget / €2,700,000

Padua Train Station Square is one of the largest squares in front of the existing train station and in the adjacent urban spaces. Every year, 18.5 million people cross the square. The physical transformation enabled an effective development of a larger urban space. Badly blended to the city, it used to be characterized by unrelated contingent architectural elements and traffic conflicts. Most parts of the square were occupied by transportation platforms (bus, tramway and taxi), and there were no organized paths, the cyclist and pedestrian flows were hard to navigate, and bicycles were randomly parked. The project therefore required a more comprehensive rethinking of the features of a contemporary multi-modal hub, with a high-quality organizational flow.

The project aimed to create a high-quality, urban space available for use by the community at a reasonable cost, providing mobility, comfort, and security. It is designed to easily direct the different users, changing the existing geometry and restoring its size, and reducing automobile circulation and roads, thereby increasing walking and cycling paths to promote slow-mobility infrastructure.

The square is differentiated into three main areas with distinct characters sqd uses: the western part is planted with trees for providing shade, the central part is monumental and dedicated to internal circulation, and the eastern part is installed with temporary steel structures for bicycle parking. The design offers continuity

▬ New pavilions	● Multi-modal	1 Car parking	7 New bicycle path
● Slow mobility	pedestrian path	2 Tram stop	8 New pedestrian area
pedestrian path	bicycle path	3 Train station	9 Tram line
bicycle path	authorized safety vehicles	4 Train station square	10 Bus line
● Vehicular mobility	taxi	5 Bicycle path	11 Taxi line
		6 Bicipark	12 City center

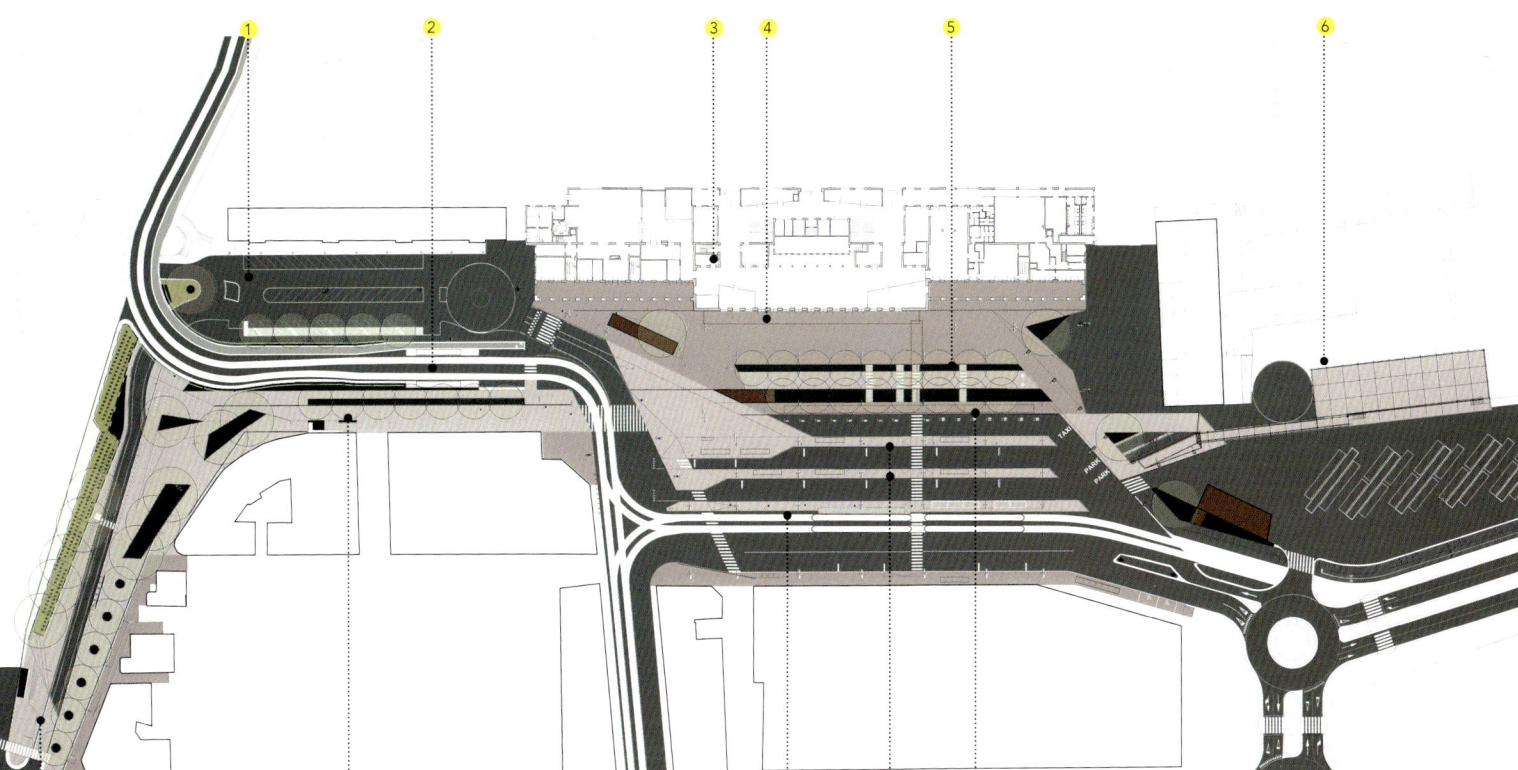

Master plan

01) General view of Padua Train Station Square
02) Top view of the pavilions in Corten
03) Aerial view of Padua Train Station Square and pavilions

between related parts, and orientation to the different users through vehicular, pedestrian, and bicycle paths. The longitudinal bicycle-pedestrian path is also lined with trees. The materials and finishing of the paths and square have been selected with users' comfort taken into consideration. In order to control the flows of different modes of travel, multi-function platforms in colored concrete (bicycle and pedestrian areas and controlled traffic) are designed to be different from the asphalted areas for vehicles. With basic safety and unitary considerations, the night-light project has three different levels: a diffused light created by 26-foot (8-meter) poles for the streets and by 16-foot (5-meter) poles standing on the multi-functional areas, and a low lighting all along the cycling lanes.

The square has been increased from 328.08 feet (100 meters) to more than 984.25 feet (300 meters) in length (east and west), and from 65.16 feet (20 meters) to 104.98 feet (32 meters) in width (north and south) by reducing the number of public transportation lanes. The bike and pedestrian paths have been tripled from 1.24 acres (0.5 hectares) to nearly 3.71 acres (1.5 hectares). The square redefined in size is perceived as a continuous space, free from obstacles and rescaled through new trees and facilities such as the bicycle parking. New vegetation that expands the existing green structures has improved the natural habitat for birds, reintroducing nature into an urban context strongly characterized by mobility.

(04) The new urban space, available and useful from users: the square in the summer with new trees of Robinia pseudoacacia
(05) Detail of the topography of the square in porphyry, used by people as a seat
(06) Printed glass taxi rank shelter
(07) The shadows of the glass taxi shelter
(08) The soft-mobility system and the cycle park

Pedestrian and cycle mobility

Vehicular mobility

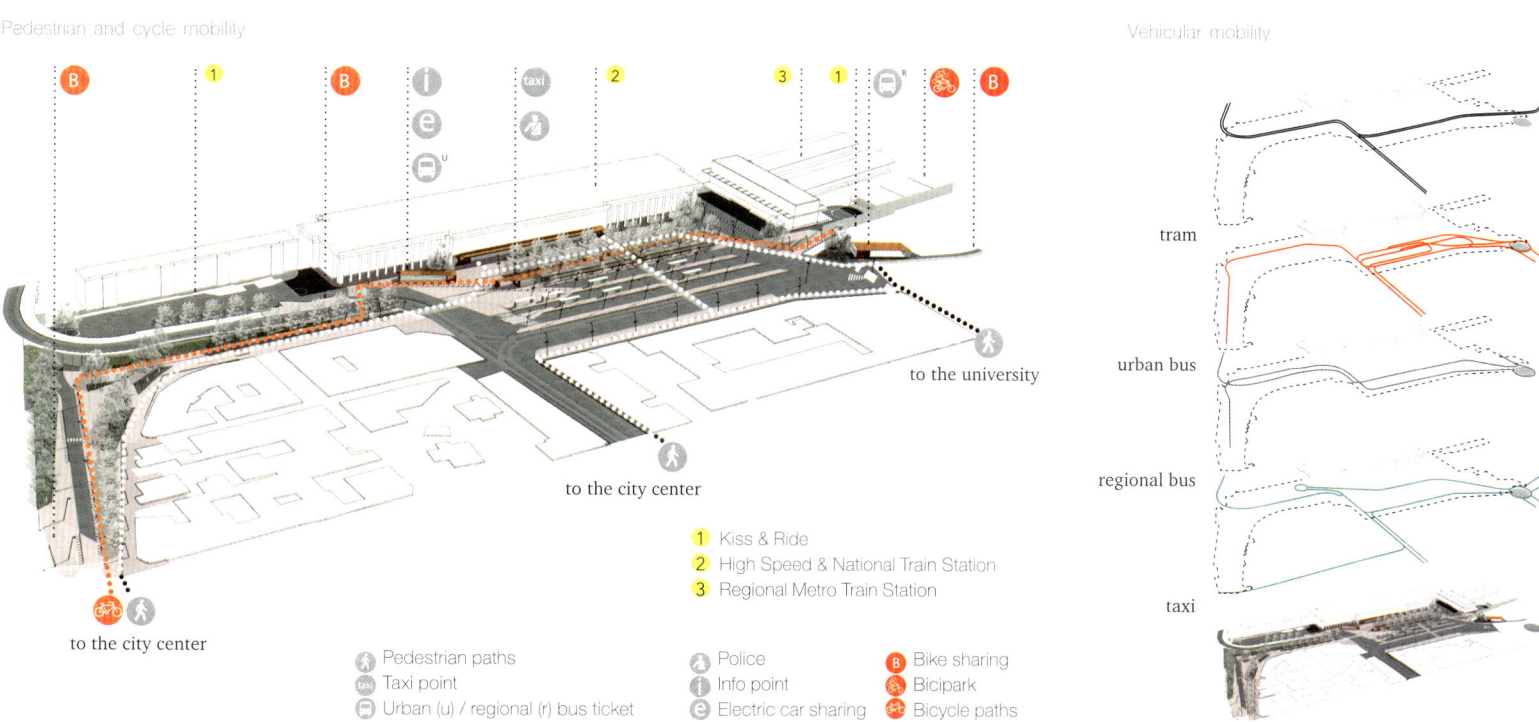

to the university

to the city center

to the city center

1. Kiss & Ride
2. High Speed & National Train Station
3. Regional Metro Train Station

- Pedestrian paths
- Taxi point
- Urban (u) / regional (r) bus ticket
- Police
- Info point
- Electric car sharing
- B Bike sharing
- Bicipark
- Bicycle paths

tram

urban bus

regional bus

taxi

109

Permeability

● black stone gravel

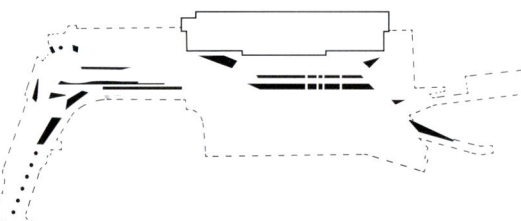

Waterproof surfaces

stone | porphiry
iron oxide pigmented concrete

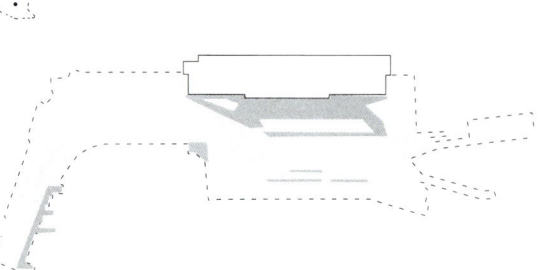

Pavement and materials

09 The covered walkway between the station and the taxi shelter
10 The multi-modal platforms made of colored concrete make the areas more versatile and control the different mobility flow
11 The square in porphyry in front of the station
12 Detail of the mineral and permeable surfaces of the square of Codalunga Street

Vegetation and surface design | ▓▓ Slow-mobility surface (stone) | ▓▓ Multi-modal surface (iron-oxide concrete)

pedestrian path
bike path

pedestrian path
bike path
authorized safety vehicles
taxi

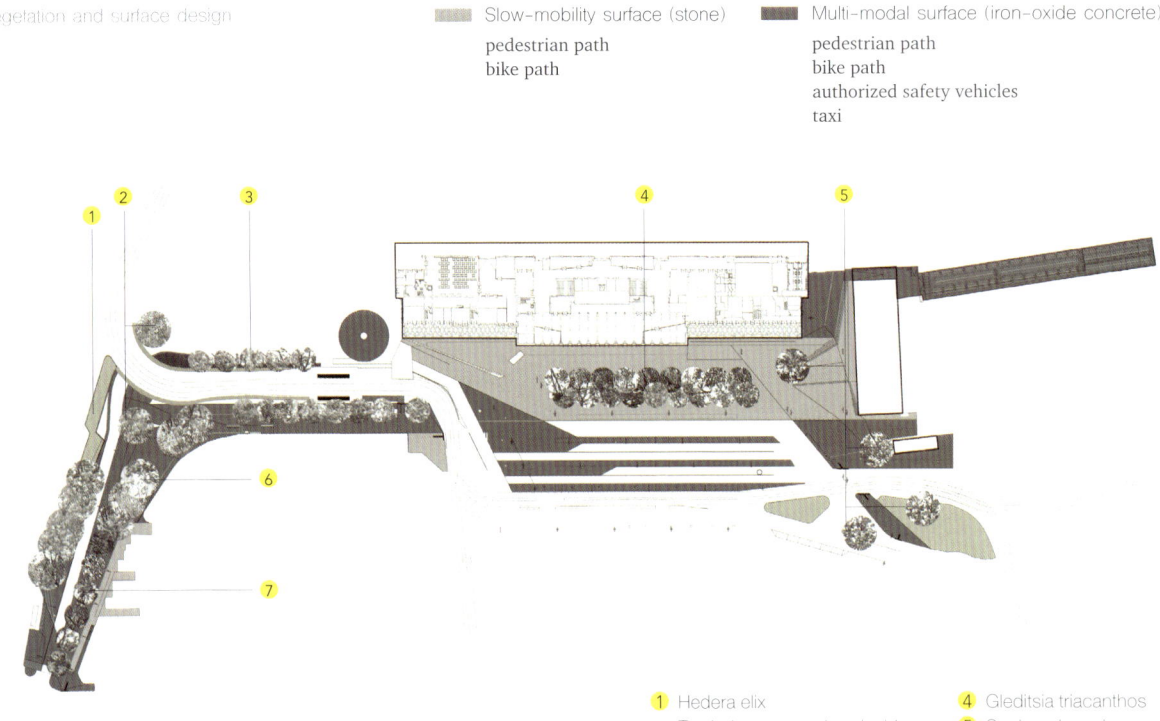

1. Hedera elix
   Trachelospermum jasminoides
2. Celtis australis
3. Robinia pseudoacacia "Bessoniana"
4. Gleditsia triacanthos
5. Sophora japonica
6. Platanus occidentalis
7. Tilia cordata

(13) Detail of topography in Codalunga Street
(14) The soft-mobility path
(15) The concrete shape helps to mark the pedestrian flows and paths
(16) Detail of the pavement and elements for flow separation
(17) Detail of road topography in Codalunga Street
(18)–(23) Detail of the mineral and permeable surfaces and the climbing plants on the tram overpass of the square of the Codalunga Street

Piazza Nember

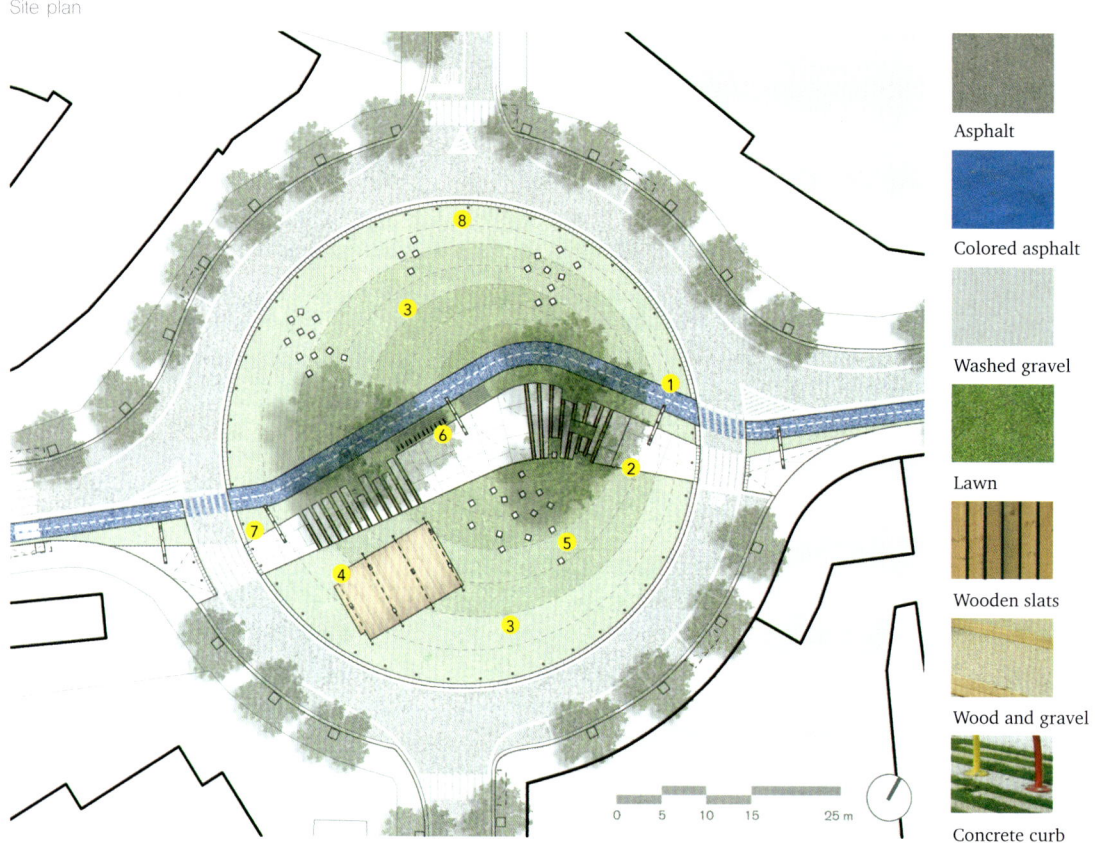

Site plan

Asphalt
Colored asphalt
Washed gravel
Lawn
Wooden slats
Wood and gravel
Concrete curb

1. Bicycle path
2. Pedestrian area
3. Grass lawn
4. Wooden stage
5. Seats
6. Bicycle rack
7. Street lamp
8. Parking barrier

● Location / Jesolo, Italy ● Area / 21,344 square feet (2000 square meters) ● Completion / 2012 ● Landscape design / Valeri.Zoia Architetti Associate, Stradivarie Architetti Associati ● Photographer / Simone Zoia, Giampietro Zoia ● Client / Municipality of Jesolo, Italy ● Budget / €180,000

Piazza Nember is not "square" either in form or function; the arrangement of its spaces seems dedicated to the infrastructure. It is a large traffic circle surrounding a large central green space with no specific landscape design. Previously, this circular path was characterized by a common lamellar wood structure, a poor lighting system, and some basic street furniture. The sidewalks along the commercial activities and hotels facing the square were made of poor materials and were in a blighted state. Moreover, the section of road became unnecessarily wide (about 40 feet [12.2 meters]) and overstated after the track road had been changed into a one-way road.

This task provided a re-qualification of Piazza Nember, with the intention of providing spaciousness and a functional organization more in accordance with a square, while dealing with the connection, both in a visual and functional way, of the two streets (Mille Street and Verdi Street) through a walking and cycling link. The project was developed around the main elements of the green area, which characterizes the existing traffic circle. In particular, the existing trees became the main references for the project. The project has kept the two "Quercus ilex" groups, characterized by high-quality fronds. To maintain the "naturalistic" aspect of the project, a slightly asymmetrical treatment of the ground was adopted, achieving the realization of the pedestrian and cyclist crossing without harming the roots of the existing plants, and lifting the ground from flat to a slight slope (2.5 percent to 4 percent) to a maximum height of 31.5 inches (80 centimeters) on the edge. The walking path was built with the well-known washed-gravel technique for parks and gardens, using a mixture of white concrete and white small-sized aggregate. The paving next to plants has been made with attention to details, leaving a reasonable distance from the plants themselves. The designer chose to have the pedestrian crossing pass through the groups of Quercus ilex, in order to let the passer-by perceive the aesthetic quality and create a shaded stopping point. A blue asphalt two-way bike path now runs parallel to the walking path. The green area is now equipped with some single seats placed in clusters, creating some micro settings.

01. Top view of the roundabout: the bike path passes through the green central space
02. The continuity of the bike path beyond the specific square area

Section

According to the ready-made principle, the wooden board has been replaced by leaning it against a zinc-coated iron structure, which is removable for events. The green-square context has been set in safety, as regards the perimeter roads, with a road dissuasive system. The road has been reduced and asphalted with a sowing of white limestone. This solution aims to stress differences from the asphalt in Dei Mille Street and Verdi Street, displaying an improved public space and the predominance of pedestrians over cars. In order to make it easier to access, the project added an augmentation of the degree of bending, both in the traffic circle entry and exit. All the horizontal and vertical street signals have been changed, too.

The illumination system is arranged into four main elements. For roads, the existing lighting has been preserved. For walking and cycling tracks, five punctual double-height elements have been placed to light up the two tracks at the same time. For green areas, the lighting system focuses only on the single seating places. The stage lighting system is defined by a white floodlighting device.

(03) The central area of the infrastructure becomes the living space for socialization

(04) The new public space is emphasized by a change of height from the street level

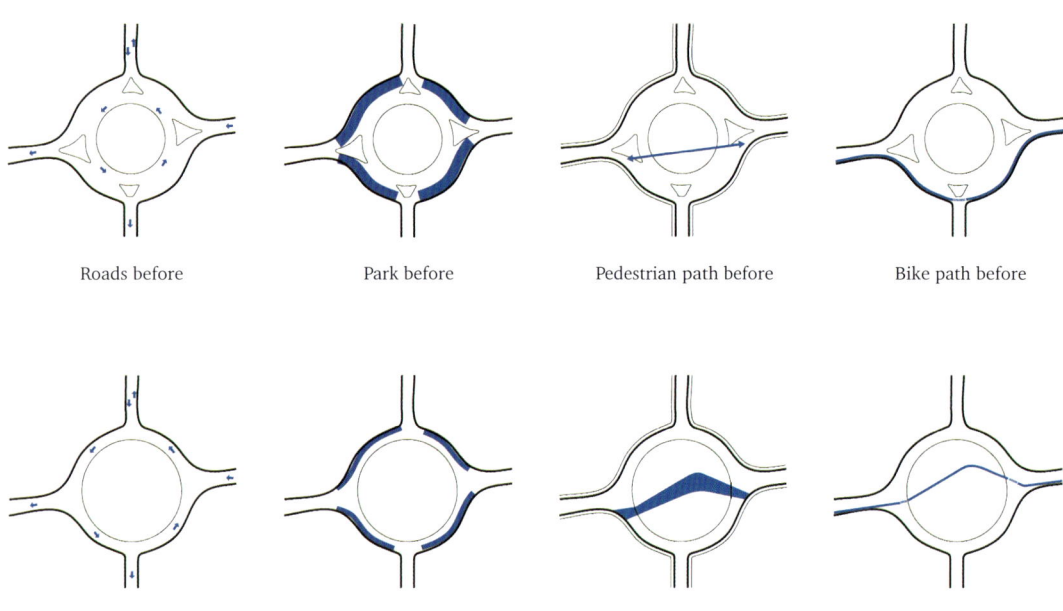

Diagrams of circulation comparison

Roads before — Park before — Pedestrian path before — Bike path before

Roads after — Park after — Pedestrian path after — Bike path after

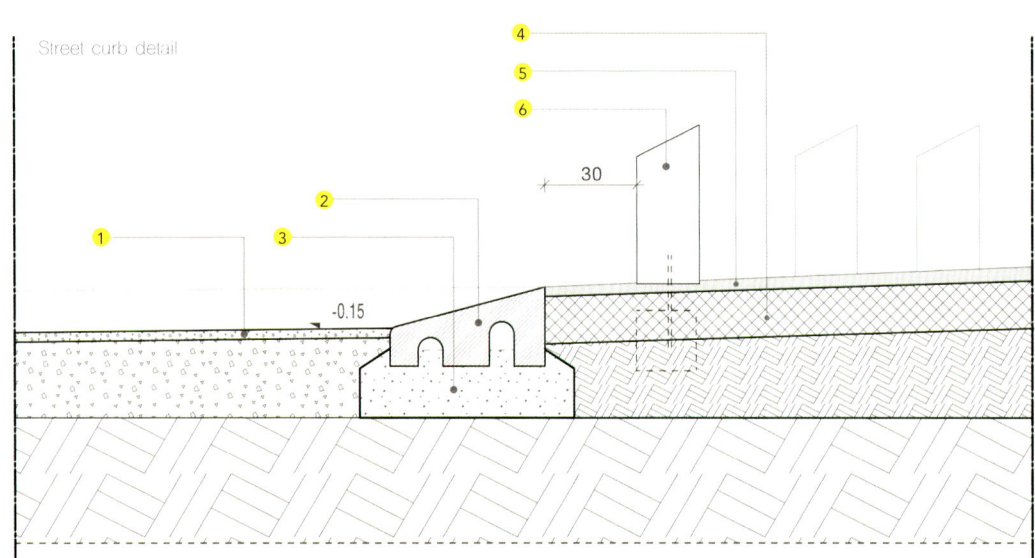

Street curb detail

1 Colored asphalt pedestrian path
2 Concrete curb
3 Concrete basement
4 Excavated land
5 Dirt and grass
6 Street bollard

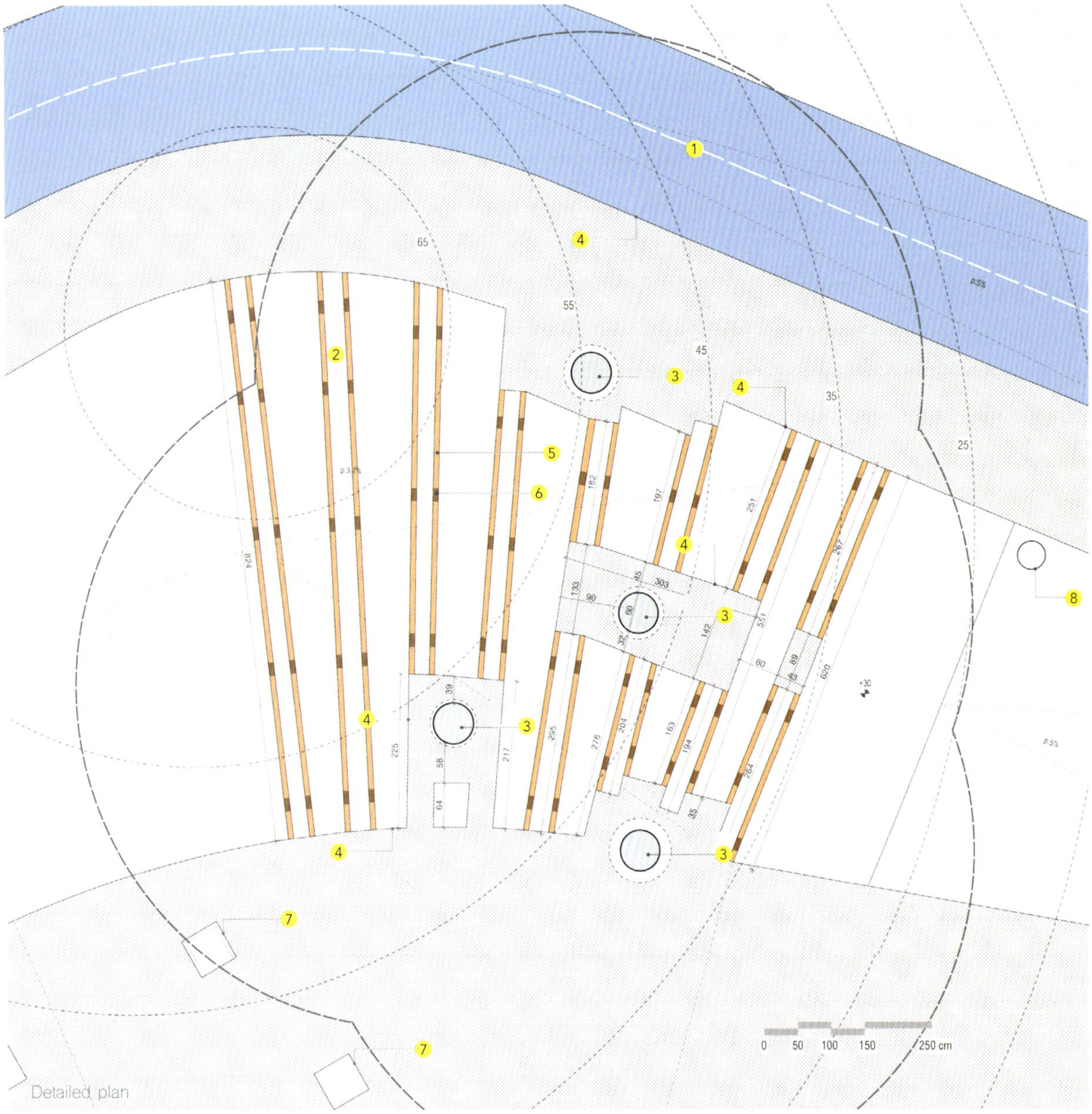

1. Colored asphalt bicycle path
2. Washed gravel pedestrian path
3. Existing Querus ilex tree
4. Metal bar
5. Wooden slats treated for outdoor use
6. Channel drainage mixed gravel
7. Metal seat
8. Litter bins

Detailed plan

05. The pedestrian and bike paths run parallel courses, separated by a grass pad
06. The in-between green space is reserved for the parking of bikes
07.-08. The pedestrian and bike paths interplay with the existing trees, creating specific design of the tracks

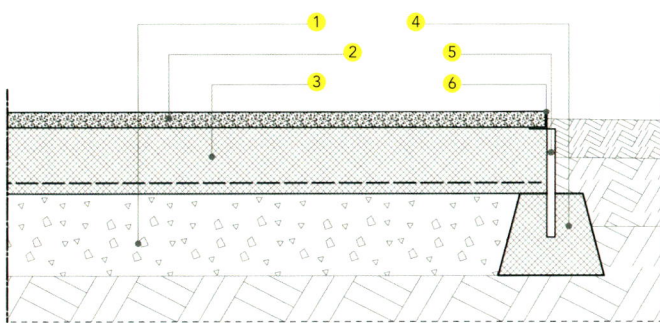
Washed gravel detail of pedestrian path

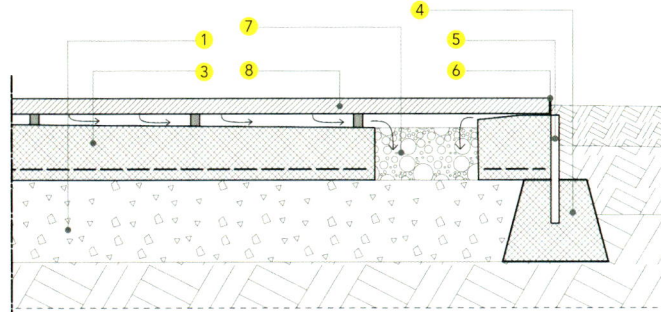
Wooden slats detail of pedestrian path

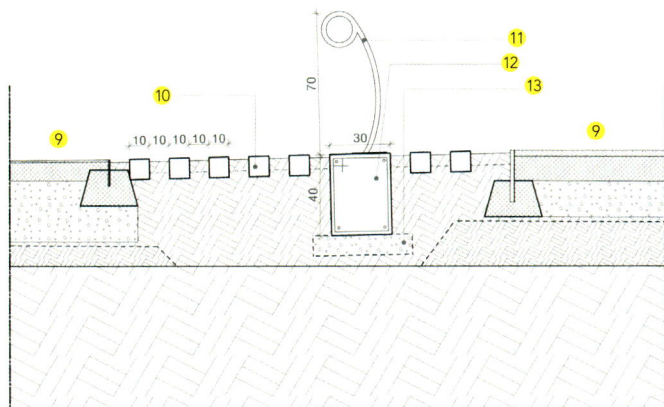
Bicycle stand detail

1. Mixed gravel basement
2. Washed gravel surface
3. Concrete screed
4. Concrete basement
5. Wooden slat
6. Metal bar
7. Bike path
8. Wooden slats
9. Path
10. Concrete curb
11. Colored metal bicycle stands
12. Concerte basement
13. Mixed gravel base level

Adolf Horn Avenue

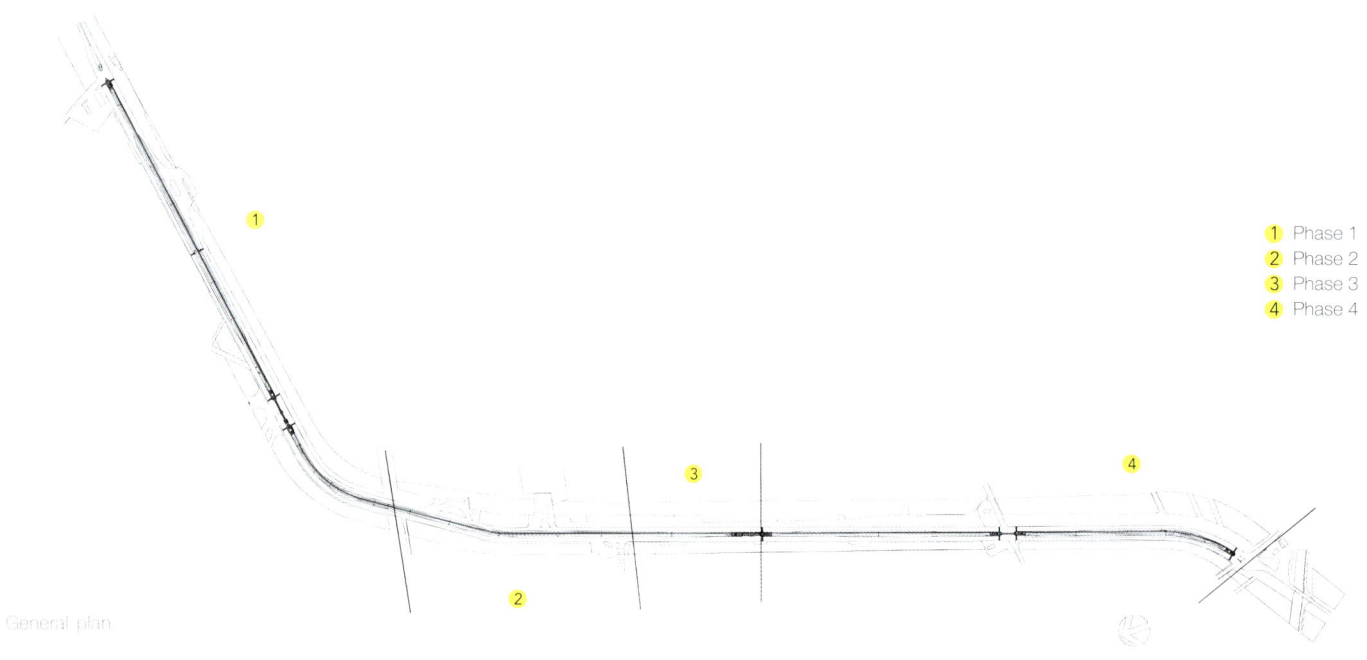

General plan

1 Phase 1
2 Phase 2
3 Phase 3
4 Phase 4

● Location / Tlajomulco de Zuñiga, Mexico ● Length / 1.9 miles (3.1 kilometers) ● Completion / 2011 ● Landscape design / AGRAZ Arquitectos ● Photographer / Martin Opladen ● Client / Tlajomulco City Government ● Budget / US$2,150,000

As in many other parts of the Guadalajara Metropolitan Area, Adolf Horn Avenue was built to connect diverse regions within Tlajomulco de Zuñiga city. It brought enormous benefits for thousands of people for whom driving was the priority. Nevertheless, the great central space of the avenue was omitted from any urban or architectural project, becoming a large wasteland over time. Therefore, this urban remodeling project, which aimed to optimize the road usage, came up with a bold proposal in the reassessment of the median strip.

The initial government assignment asked for a linear park, requiring more green areas within the central space. The first obstacle was budgetary because the remodeling project needed to build a main sewer to solve the problem of flooding during the rainy season. This lead to changing the project so that, instead of a linear park, the median strip would focus on alternative transportation and non-motorized mobility, with a 1.86-mile (3-kilometer) bicycle lane and a long garden. The Adolf Horn Avenue project was developed out of three concepts: park areas at the crossings of pedestrian zones with small plazas for people to use for lunch or social gatherings; a bicycle lane and a promenade for transportation and sports; and a line forest of 2000 trees in a single species.

It was a long-term project that depended on maintenance, especially on its first years. It meant giving priority to public space and green areas, over automobiles. The new central strip of Adolf Horn Avenue showed how much more roads with public space and nature are needed. What's highly significant is the number of ash trees planted on the 1.86-mile (3-kilometer) strip of the eastern part of the Metropolitan area, as well as the shared commitment of Tlaquepaque to continue the work on the section of its property within Tlajomulco.

① Initial milestone as a landmark
② Intersection plaza for social gatherings
③ Entrance of bicycle and pedestrian lanes from the intersection plaza

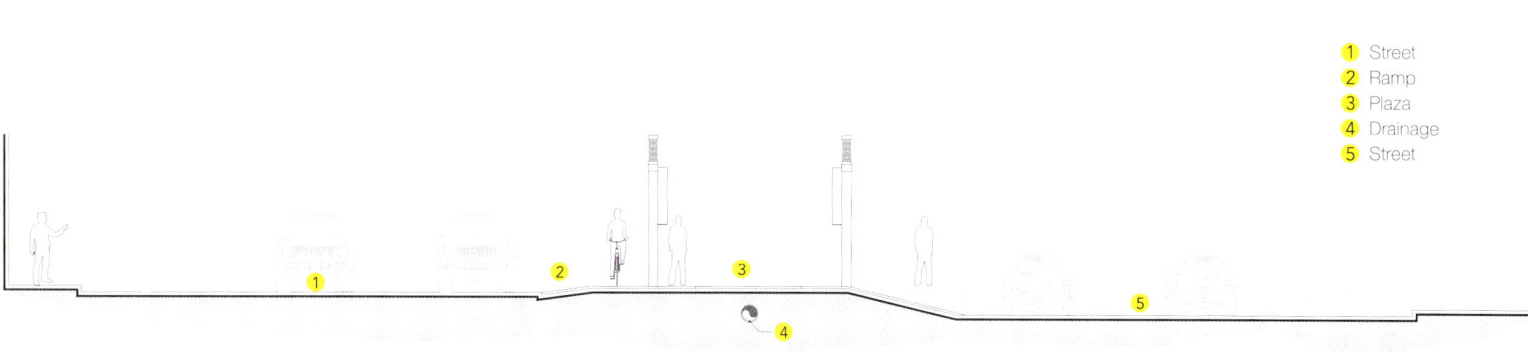

1 Street
2 Ramp
3 Plaza
4 Drainage
5 Street

Section 1

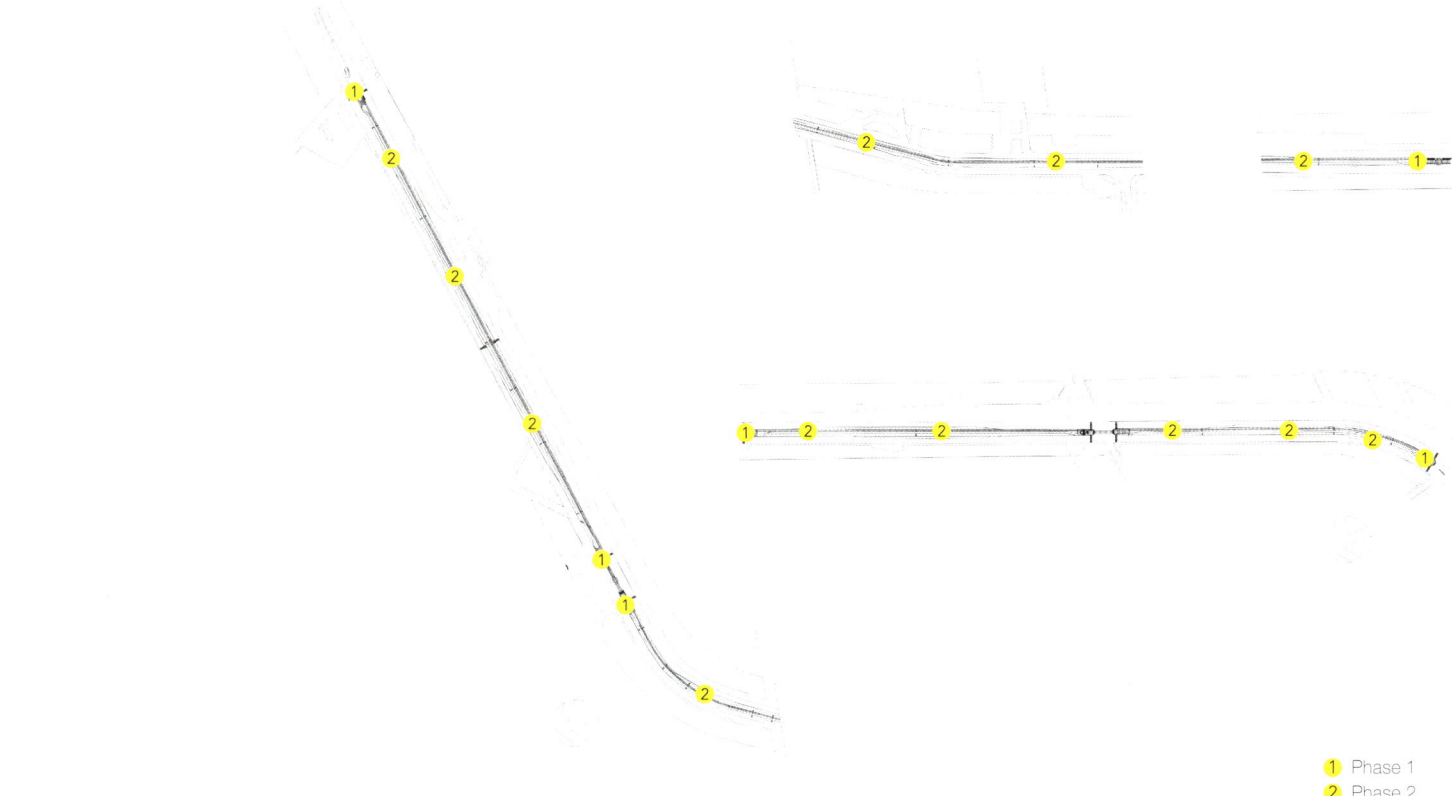

1 Phase 1
2 Phase 2

(04) Exclusive bike path signage for the bicycle lane
(05) Seating area
(06) Intersection of bicycle and pedestrian lanes with the road

1. Street
2. Ramp
3. Bikeway
4. Sidewalk
5. Sculpture
6. Drainage

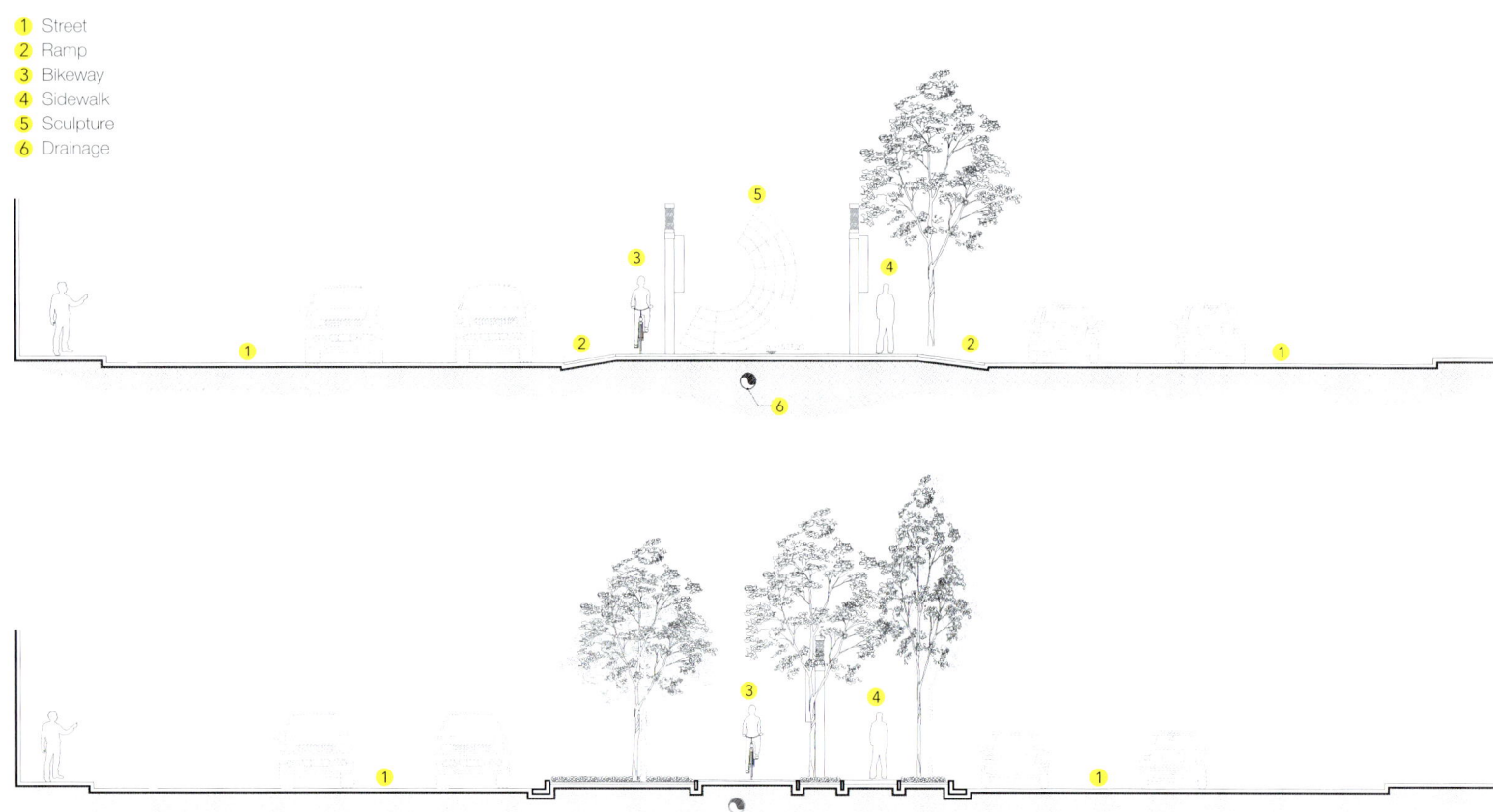

Section 2

07 - 08 Accessibility detail of the intersection plaza
09 Pedestrian sidewalk
10 Two-way bike lane

Bourke Street cycleway

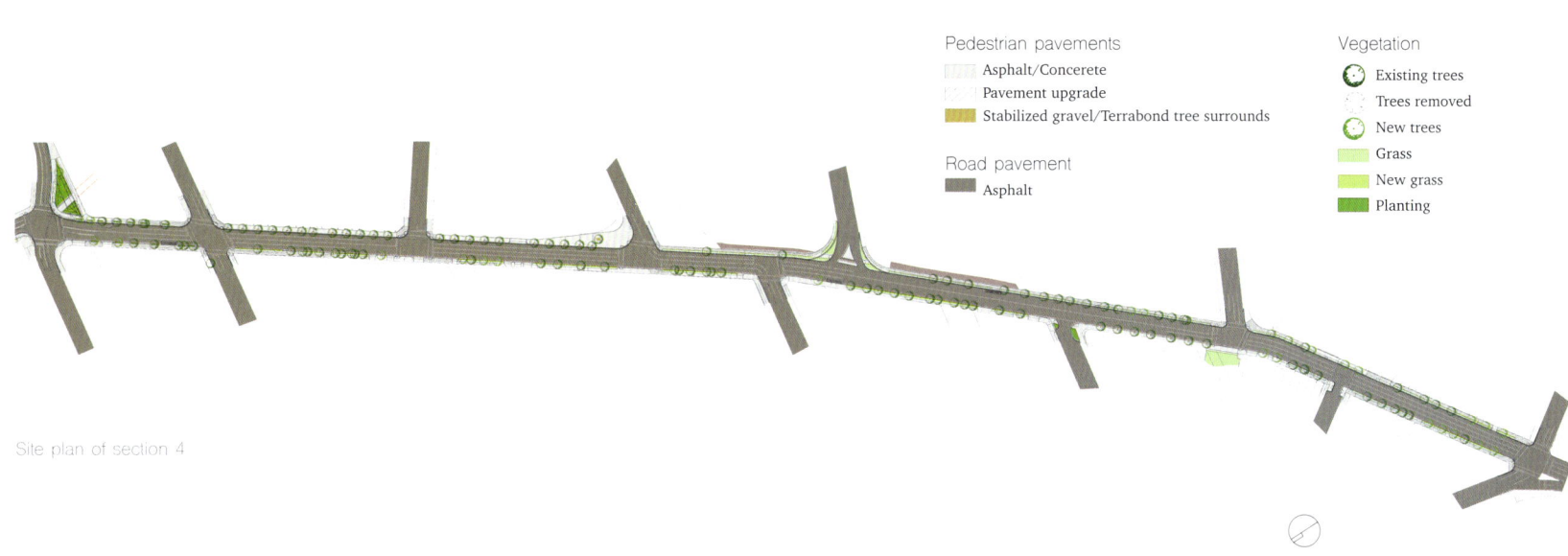

Site plan of section 4

- Location / **Sydney, Australia** - Length / **1.99 miles (3.2 kilometers)**
- Completion / **2012** - Landscape design / **Group GSA** - Photographer / **Simon Wood** - Client / **City of Sydney** - Budget / **AU$24,000,000**

Bourke Street is the first large-scale two-way separated bike path in Sydney. It is a part of the City of Sydney's cycling strategy to increase the quality and scope of the cycle network. The 1.99-mile (3.2-kilometer) project is an upgrade of the existing on-street cycling route, providing improved safety and amenities for cyclists, motorists, and pedestrians by increasing separation from vehicle traffic.

The bike path adopts a suite of separated profiles, facility typologies, and intersection solutions for the varying sections of the road. The two-way bike path scheme is a solution to the problem of retrofitting separated facilities within Sydney's relatively narrow street profile. By harvesting excess space out of the road corridor, a safe space can be created for cycling that retains parking on both sides of the street in most circumstances. The profile subtly adjusts its form along the length of Bourke Street to ensure that the existing street amenities are enhanced and maintained. The route includes shared paths to ensure continuous connections for cyclists, while responding to the changes in the urban context and opportunity.

The resulting complexity at intersections has been resolved with a hierarchy of treatments. Minor laneways have been converted to driveway crossovers and paved intersections that create a shared environment with the pedestrians as the priority. More significant intersections with no signal integrate marked pedestrian crossings with bike slip lanes that result in continued cycling priority across these intersections and safe places for cars to wait while giving way. Finally, intersections with signals provide periodic priority with separate signals for bikes. These intersections required extensive modeling of the traffic functions to achieve project approval.

The selection of materials, pavements, and plantings vary along the corridor to respond to the character and condition of different precincts. Extensive modification to stormwater infrastructure was used to treat water in rain gardens.

The finished facility is an elegant solution that delivers multiple benefits and respects the urban context. It forms a safe, convenient, and sustainable cycling route linking into a larger network, thus reducing road congestion, cutting emissions, and improving public health. The network has improved connections between employment, recreation, and residential destinations, making cycling an attractive transportation choice. The number of riders on the Bourke Street route has increased by 250 percent. Through careful consideration of Bourke Street's built and landscape heritage, the facility also provides a new appreciation of one of the great streets in Sydney's inner east.

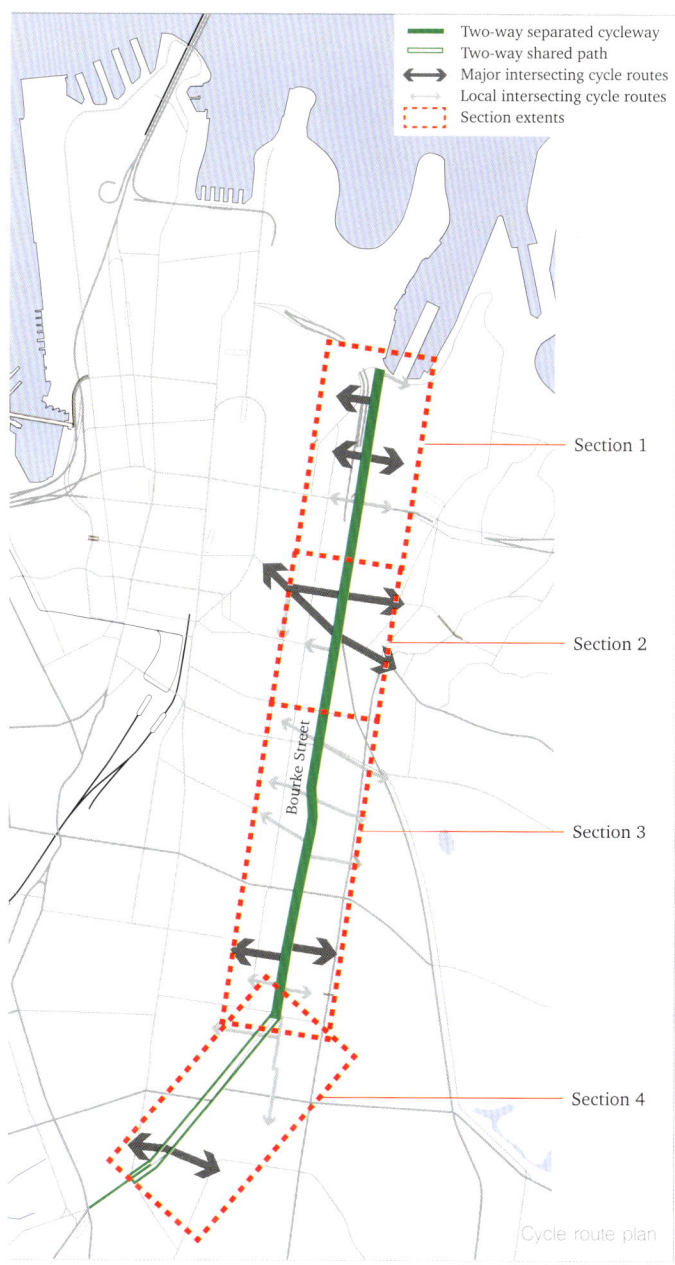

01 Separated cycleway
02 In-lane bus-stop adjacent cycleway

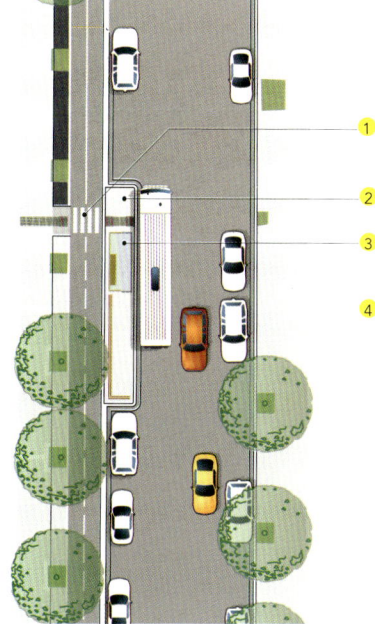

1 Marked pedestrian crossing at cycleway
2 Raised platform
3 Slim-line bus shelter
4 Bus set down/pick up in travel lane

Bus stop platform indicative plan

127

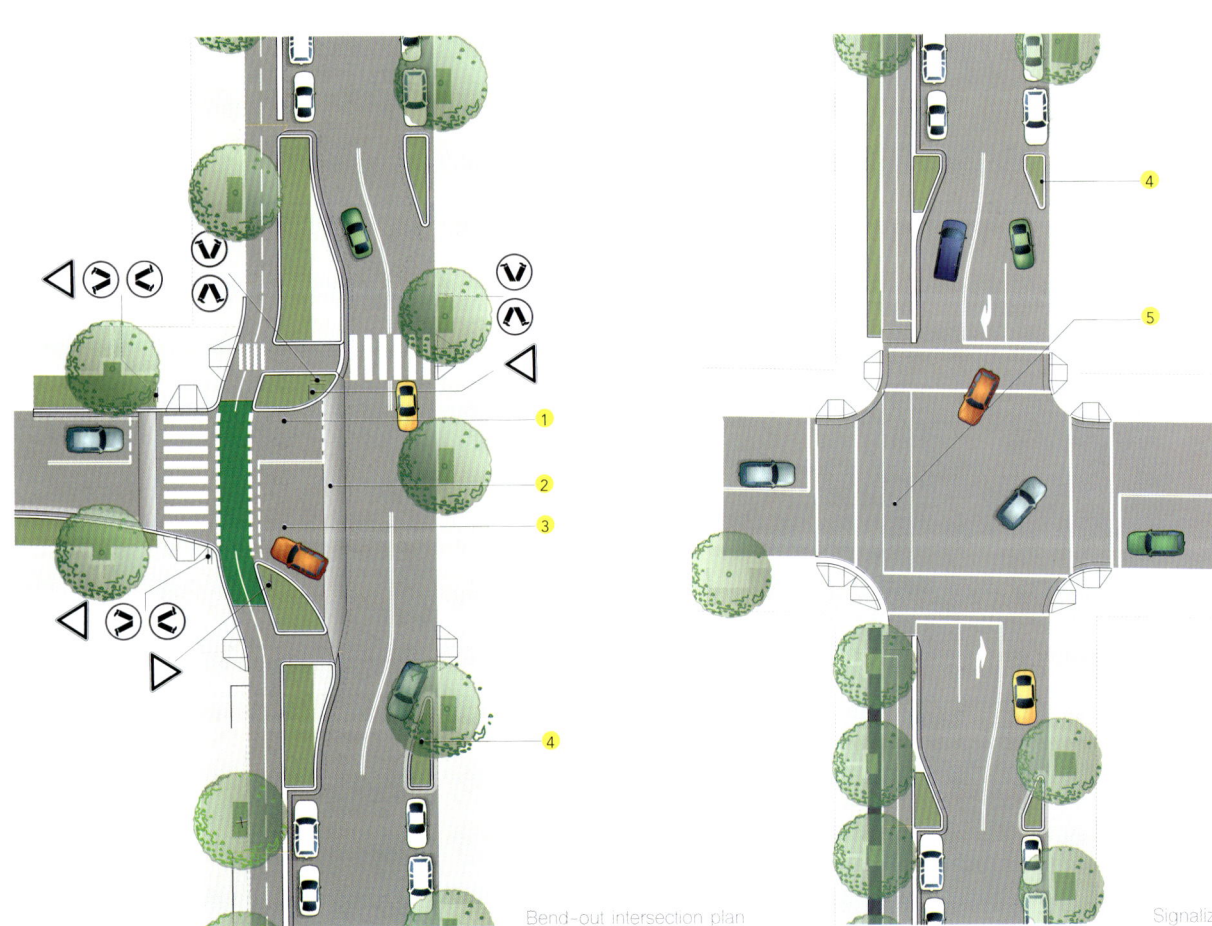

Bend-out intersection plan

Signalized intersection plan

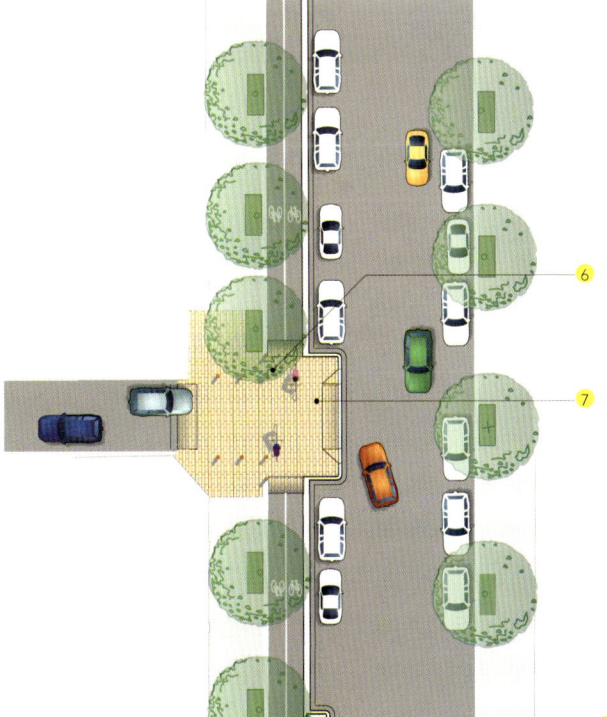

03 Bend at intersection
04 Rain garden integrated into streetscape
05 Shared environment intersection

1 Raised platform
2 Concrete ramp
3 Setback to allow cars to turn to give way to cyclists
4 Guidance barrier or line marking
5 Cycle lane
6 Bollards
7 Raised platform at footpath level

Shared environment plan

06 Bourke Street is a "Green Boulevard"
07 Planted sections separate the cycleway
08 Cycleway with rain gardens adjacent

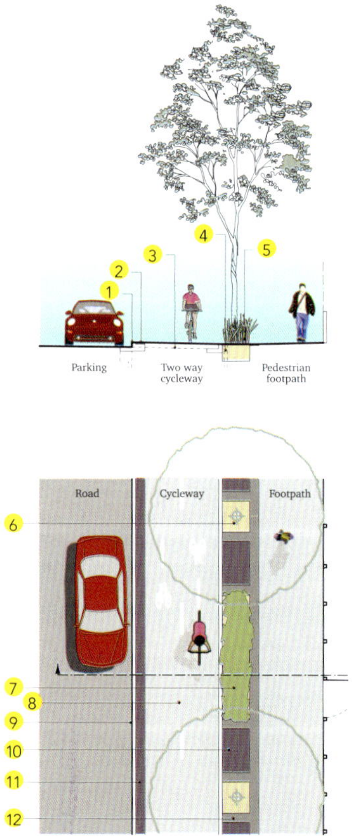

1 New curb line
2 Textured surface
3 Exisiting roadway to be covered with new asphalt
4 Stormwater drain
5 Planting
6 Tree with gravel surrounds
7 Planting
8 Two-way cycleway
9 New curb line
10 Contrast paving and planted buffer zone
11 Textured surface
12 Unit pavers
13 Existing gutter lowered by 1.97 inches (5 centimeters) to allow asphalt and curb
14 Double-stepped two-way cycleway
15 Wide median
16 Planted median
17 Tree with improved gravel surrounds
18 All street furniture to be located away from shared-path area

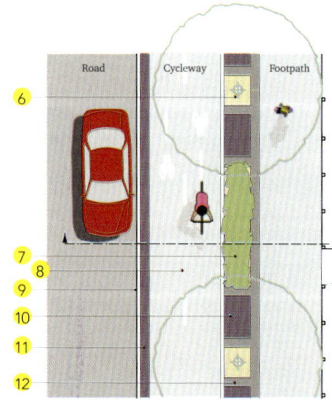

Flush with footpath and two-way separated cycleway indicative plan and section

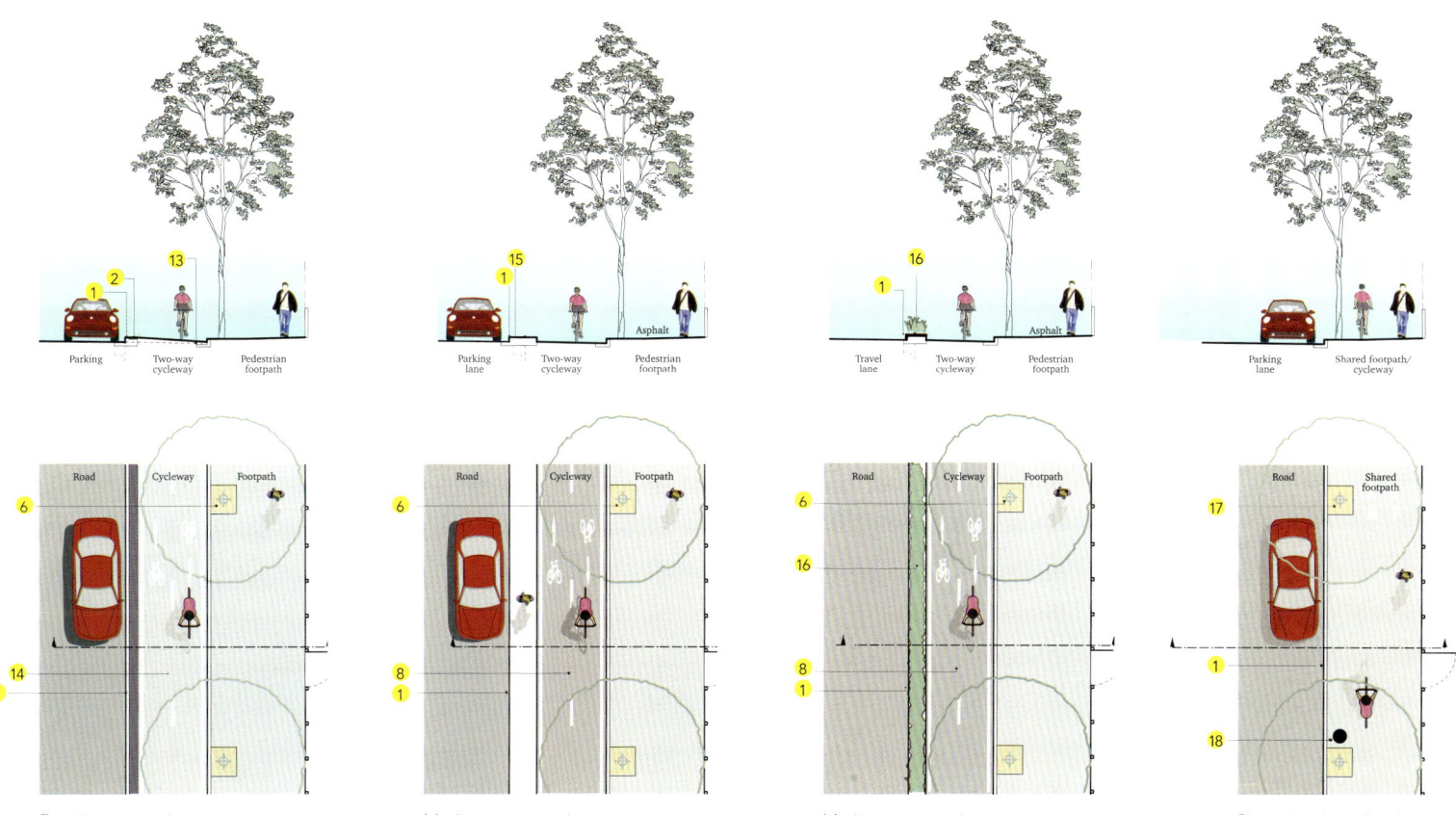

Double-stepped two-way separated cycleway: indicative plan and section

Median-separated two-way cycleway with parking adjacent: indicative plan and section

Median separated two-way cycleway with no parking adjacent: indicative plan and section

Shared path: indicative plan

George Street cycleway

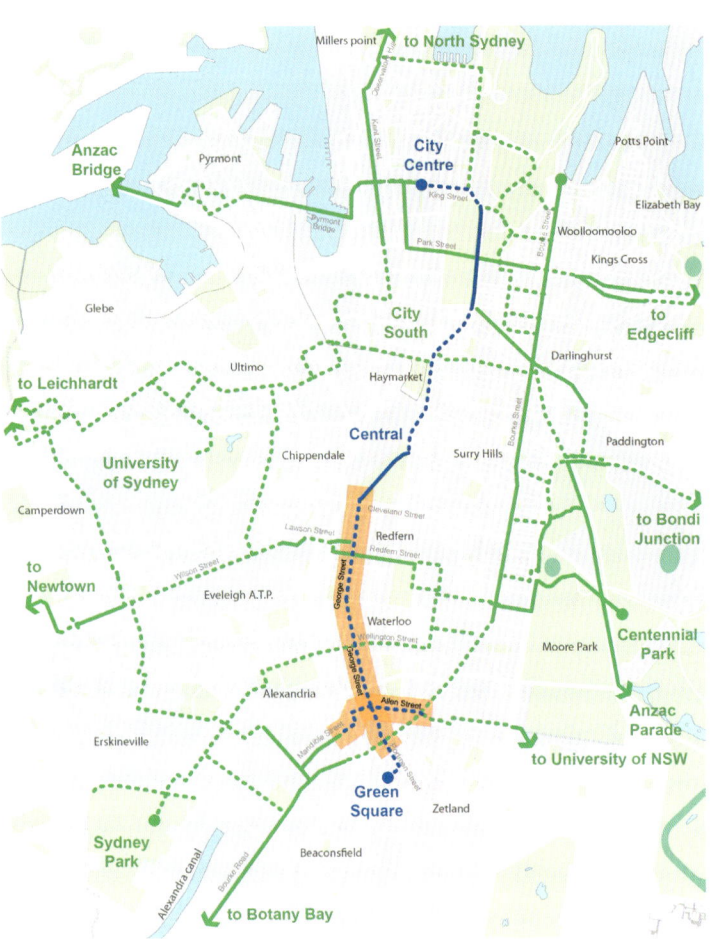

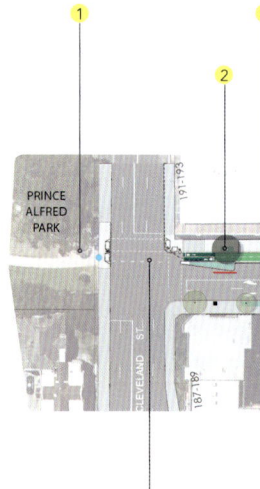

● Location / **Sydney, Australia** ● Length / **1.55 miles (2.5 kilometers)**
● Completion / **2014** ● Landscape design / **Group GSA** ● Photographer/
Simon Wood ● Client / **City of Sydney** ● Budget / **AU$12,000,000**

The George Street cycleway was originally part of a wider study of key cycle routes in the *City of Sydney Cycle Strategy and Action Plan 2007–2017*. The strategy identifies the city's priority cycling network and proposes a high level of provision of amenities on key routes through the local government area. The 1.99-mile (2.5-kilometer) George Street route provides a significant north–south link in the regional bike network from the Sydney CBD starting at Prince Alfred Park in Redfern through Waterloo and Alexandria, across Bourke Street to Green Square. The connection to Green Square is one of the key strategic moves for the project. It is estimated that the Green Square precinct is used by more than 60,000 people. The establishment of the bike path infrastructure within the urban landscape will help affirm cycle transportation as an integral part of the urban renewal of inner Sydney.

The bike path provides a 7.87-foot-wide (2.4-meter-wide) two-way separated facility retrofitted into the existing street profile and responds to varied conditions along its length. It includes a two-way separated facility as well as shared paths and various intersection treatments to ensure cyclists' safety. In one of the key areas of the project in Waterloo, which is undergoing significant changes in land use, the streets have been designed to be pre-adaptive to accommodate the bike path. Sections of the street have been changed to one way to accommodate the existing heavy-vehicle traffic. When the area develops into a residential area, which has lower traffic volume and lane-width requirements, the streets can revert back to two-way traffic lanes with bike paths.

The bike path is a project that has a strong multi-disciplinary approach, led and coordinated by landscape architects with a team of cycle experts, and traffic, civil, structural, and lighting engineers. It has been developed in conjunction with enhanced pedestrian amenities to achieve the primary project goal of urban connectivity along George Street. The public domain outcomes include significant greening of the streetscape with new tree planting, shrub planting, and stormwater quality improvements.

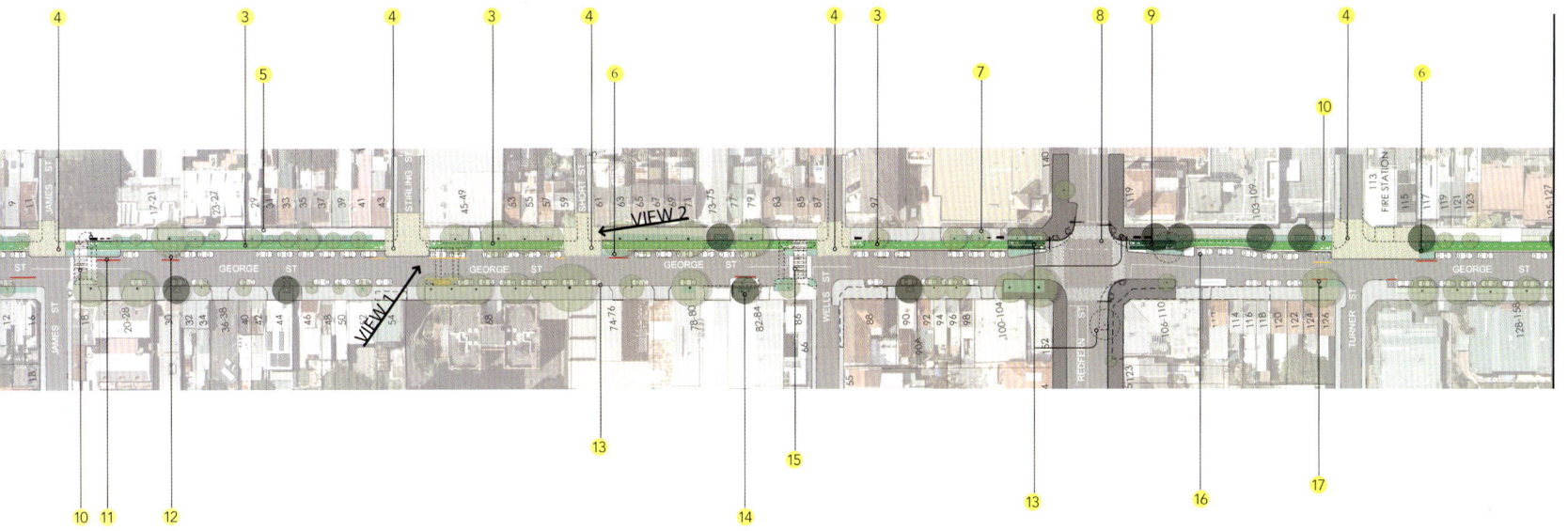

George street northern precinct site plan 1

- Existing tree removed
- Existing tree retained
- New tree
- Proposed car parking
- Car parking removed
- Car parking added
- Loading zone
- Bus stop and shelter
- Separated cycleway
- Shared path
- New footpath
- New curb ramp
- New shrub planting
- New biofiltration planting

1. Existing shared path through Prince Alfred Park to Central Station
2. Motorcycle parking relocated toward south
3. Double-stepped separated cycleway
4. Shared environment intersection
5. New footpath and curb
6. Signals modified for cycleway crossing
7. Flush to footpath separated cycleway
8. Relocated motor and bike parking
9. Signalized cycle crossing connecting to shared path in Prince Alfred Park
10. New raised pedestrian crossing linking James Street to Reconciliation Park
11. One parking space removed and motorcycle parking relocated to allow pedestrian crossing
12. One parking space removed to accommodate relocated motor cycle parking
13. Mobility parking space retained
14. Curb reduced and footpath reconstructed
15. Raised threshold replaced with new raised pedestrian crossing
16. Footpath narrowed at intersection
17. On-street bike parking

01 Shared environment intersection

02 Shared environment intersection at laneway
03 Bend-out intersection giving cyclist priority
04 Signalized intersection

Types of separated cycleway

1 New curb
2 New asphalt surface over existing rod
3 3.94-inch (10-centimeter) curb
4 New planting
5 Flush curb
6 Separated median islands
7 Curb
8 Median with planting
9 Two-way planting
10 Stepping-stone pedestrian link through planting
11 Paved pedestrian link through planting
12 Marked parking bays

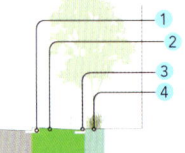
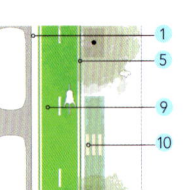
Double-stepped separated cycleway

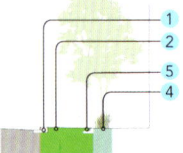
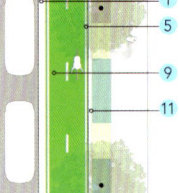
Flush to footpath separated cycleway

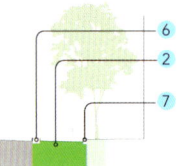
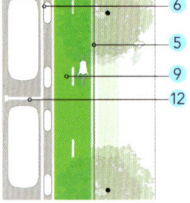
Interrupted median separated cycleway

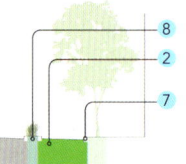
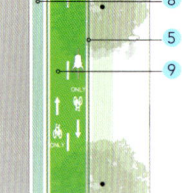
Planted median separated cycleway

George Street northern precinct site plan 2

1 Five parking spaces removed to reduce pedestrian and cycle crossing distance
2 Cyclists give way at road crossing
3 Interrupted median separated cycleway
4 Footpath reduced
5 Roundabout removed and bend-out intersection installed to allow priority cycle crossing
6 Interrupted median separated cycleway
7 New footpath and curb
8 New bend-out intersection to allow priority cycle crossing
9 Rain garden medians with biofiltration planting
10 New signalized intersection with separated cycle and pedestrian crossing
11 New motorcycle parking
12 Shared cycle and pedestrian path through road closure

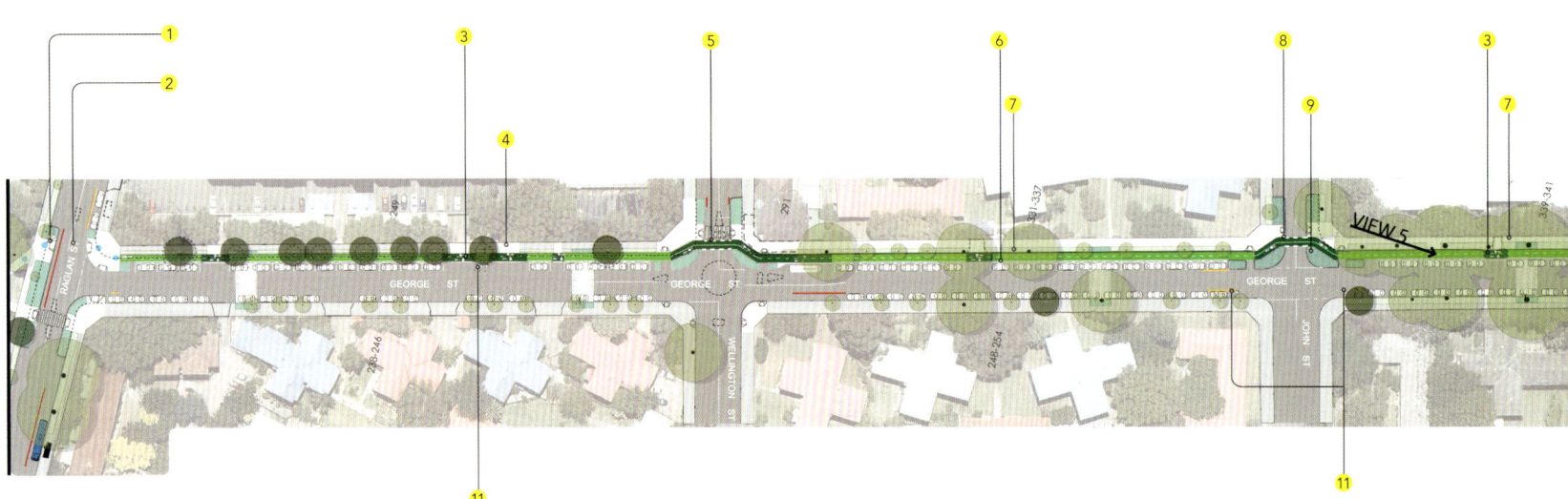

134

Cycleway intersections

a Signalized cycle crossing
b Planting area
c Priority cycle crossing at road level
d Previous curb line
e Tactile indicators
f Bollard
g Intersection raised to footpath level and paved with unit pavers
h Driveway

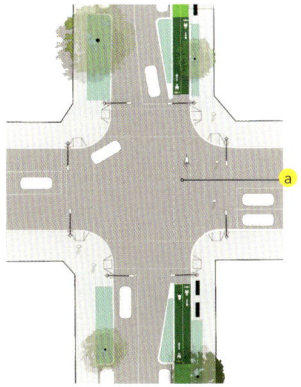

Typical signalized intersection plan

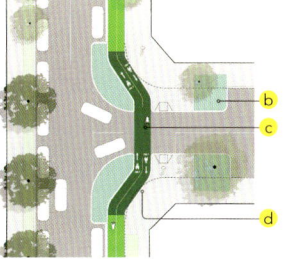

Typical bend-out intersection plan

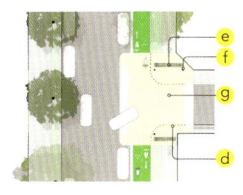

Typical shared environment intersection plan

Typical driveway plan

135

05 Shared path link
06 Separated cycleway
07 Shared path treatment

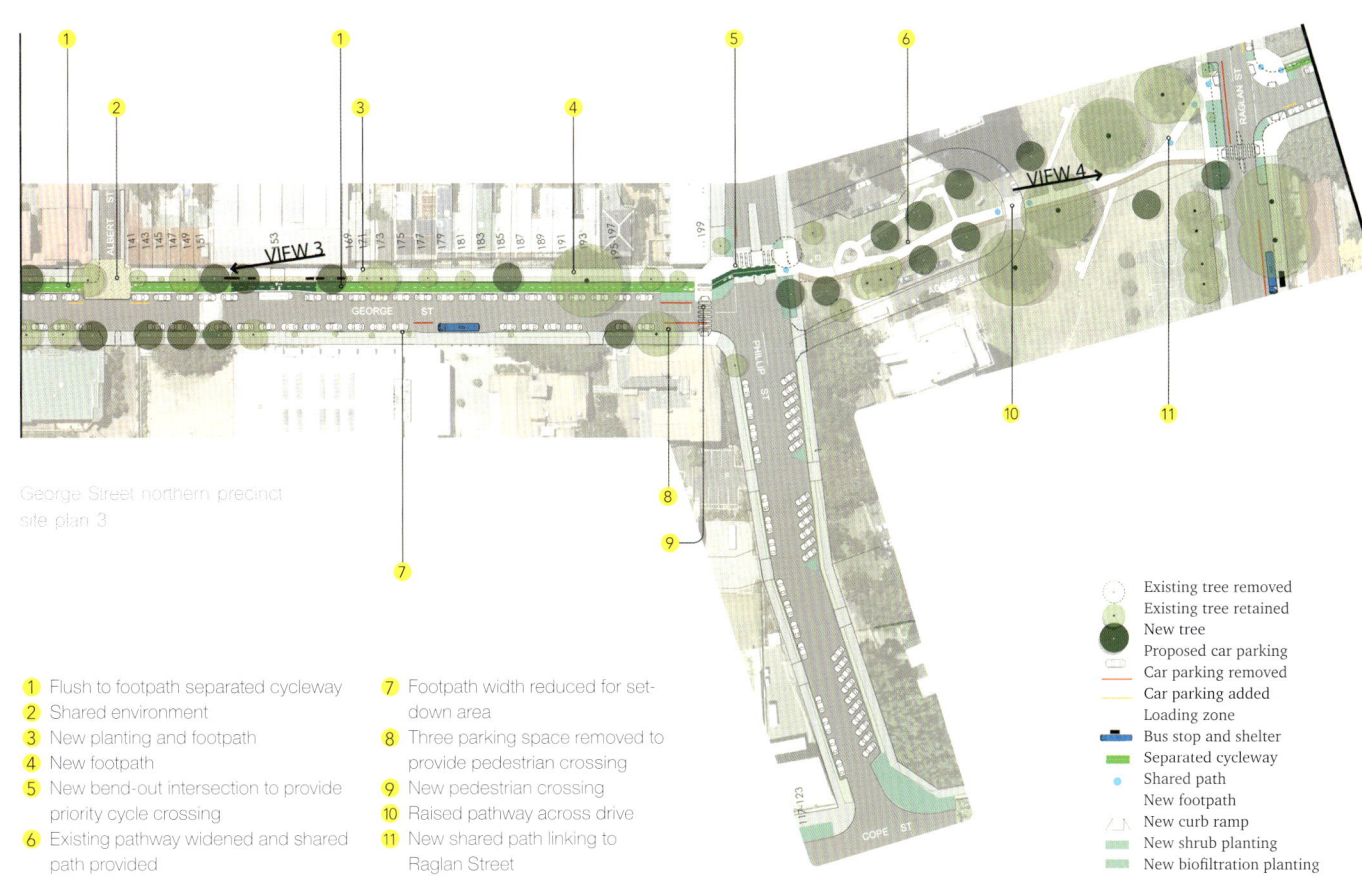

George Street northern precinct
site plan 3

1. Flush to footpath separated cycleway
2. Shared environment
3. New planting and footpath
4. New footpath
5. New bend-out intersection to provide priority cycle crossing
6. Existing pathway widened and shared path provided
7. Footpath width reduced for set-down area
8. Three parking space removed to provide pedestrian crossing
9. New pedestrian crossing
10. Raised pathway across drive
11. New shared path linking to Raglan Street

Existing tree removed
Existing tree retained
New tree
Proposed car parking
Car parking removed
Car parking added
Loading zone
Bus stop and shelter
Separated cycleway
Shared path
New footpath
New curb ramp
New shrub planting
New biofiltration planting

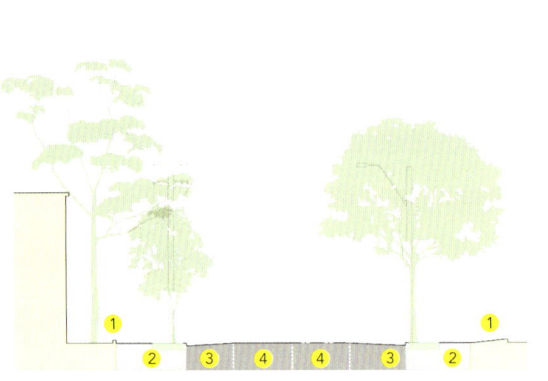

Section of existing two-way street

Section of proposed one-way street

1. Property boundary
2. Footpath
3. Parking lane
4. Travel lane
5. Interrupted median separated cycleway
6. Travel lane removed and one-way street proposed for sections of George Street

George Street northern precinct site plan 4

1. Shared path upgraded and new tree planting in park space
2. Unsignalized connection across short street
3. Interrupted median separated cycleway
4. Signals modified for cycleway crossing
5. Footpath narrowed
6. Low bump medians at driveway crossings
7. New motorcycle parking

- Existing tree removed
- Existing tree retained
- New tree
- Proposed car parking
- Car parking removed
- Car parking added
- Loading zone
- Bus stop and shelter
- Separated cycleway
- Shared path
- New footpath
- New curb ramp
- New shrub planting
- New biofiltration planting

(08) Separated cycleway with street verge shrub planting
(09) Interface of separated cycleway to shared path
(10) Separated cycleway with new trees

09

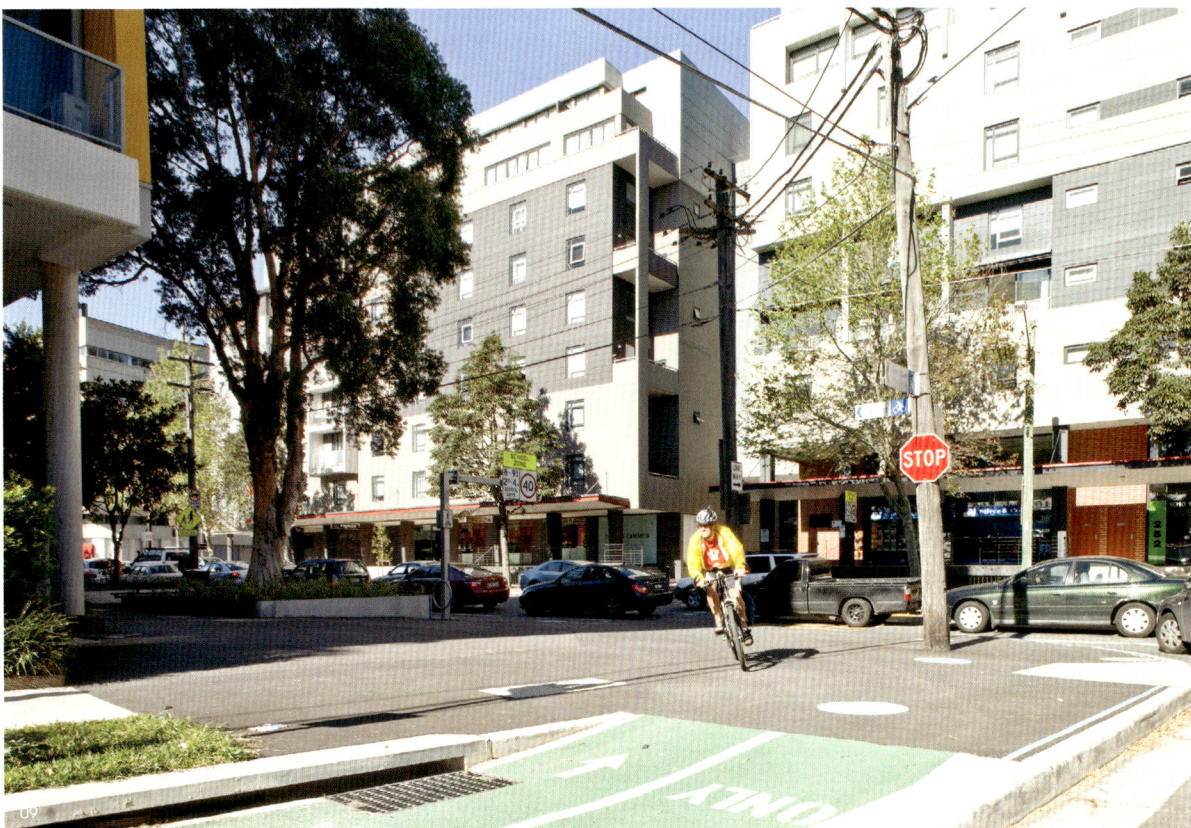

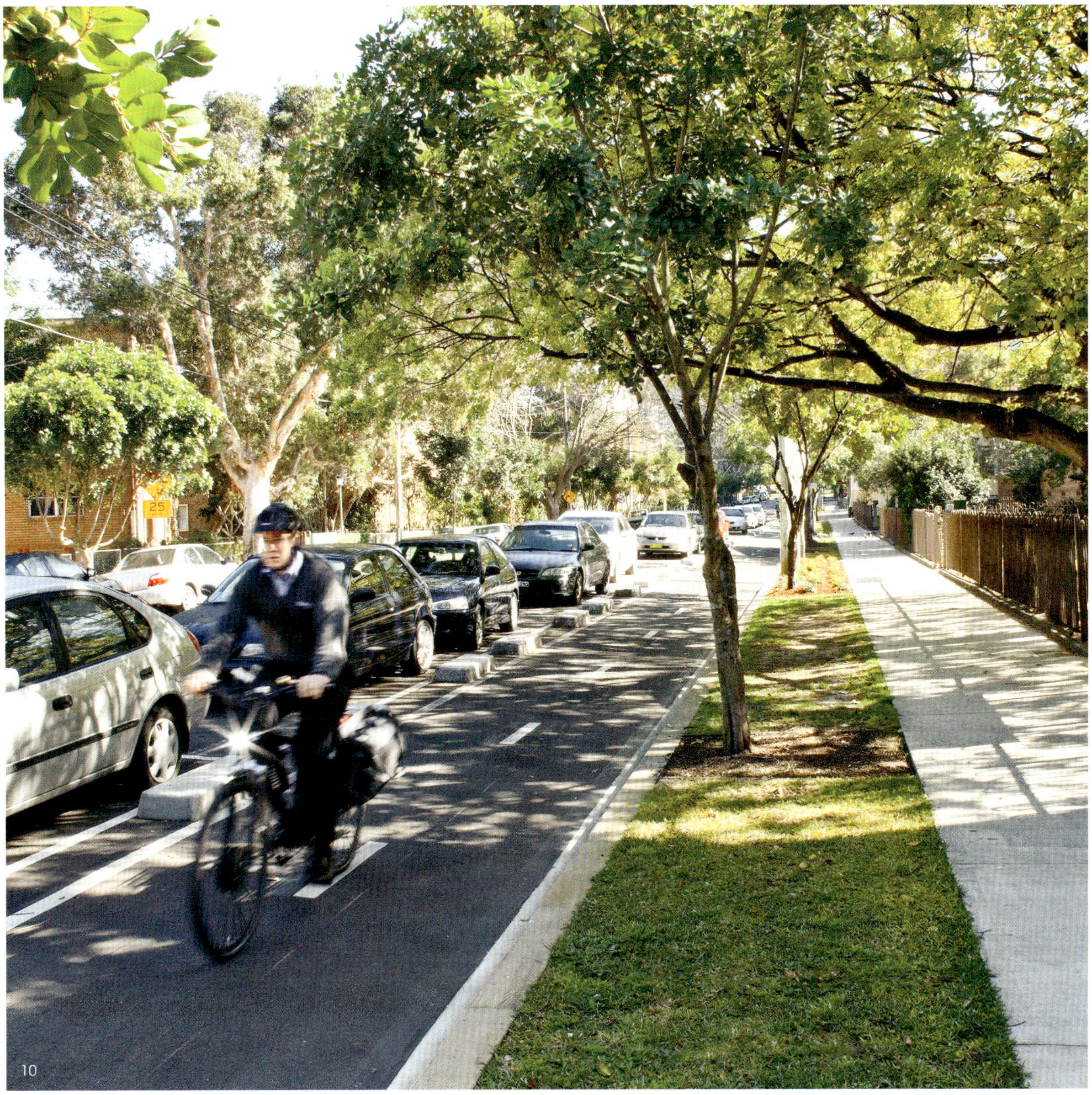

10

Queens Quay Boulevard

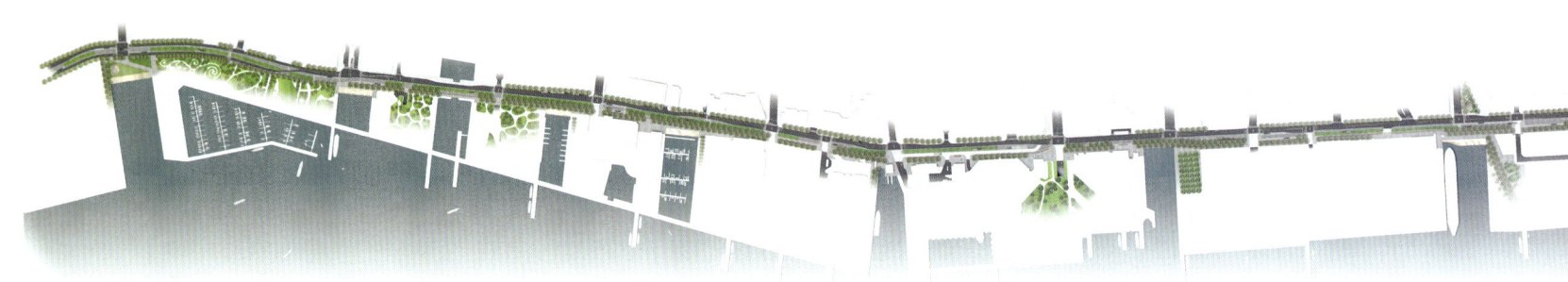

Queens Quay plan

- Location / **Toronto, Canada** ● Length / **2.2 miles (3.5 kilometers)**
- Completion / **2015** ● Landscape design/ **West 8+DTAH**
- Photographer / **Nicola Betts, West 8, Waterfront Toronto** ● Client / **Waterfront Toronto** ● Budget / **CA$129,000,000**

Queens Quay, which runs east–west in parallel to the lakefront, is Toronto's main waterfront street. The redevelopment of Queens Quay is one of the subprojects of West 8 and DTAH's award-winning Toronto Central Waterfront Master Plan.

The project has noticeably reorganized the traffic lanes: the motor-vehicle traffic lanes were reduced from four to two and shifted to the north side of the street, making room for a dedicated light rail transit (LRT) corridor in the middle and an 11.81-foot-wide (3.6-meter-wide) two-way bike lane together with a generous pedestrian promenade on the south side. The pedestrian sidewalk on the north side of street has also been renewed and widened. The bike lane is an extension of the Martin Goodman Trail, a 1.06-mile (1.7-kilometer) multi-use (pedestrian/bicycle) recreational trail along Toronto's waterfront. Before this revitalization, the bike lane ended abruptly as pedestrians entered this section of Queens Quay, forcing cyclists into lanes with faster moving motor-vehicle traffic. The new bike lane completes the missing section of the Martin Goodman Trail. During the summer of 2006, a four-story sculptural arch, a temporary landscape intervention, was constructed with more than 600 bicycles to allow the public to experience the benefits of the new proposal and to highlight the new section of the bike trail.

The Queens Quay project features high-quality, custom-designed mosaic paving, furniture, and tree-planting details, which are all inspired by the Canadian iconic maple leaf. Green- and blue-painted lines run along the whole bike lane, and blue bicycle boxes let cyclists know when they're approaching driveways, intersections, and other pedestrian crossings. White maple leaves are also painted on the trail where it approaches traffic signals. Every stone of the adjacent red and white granite pedestrian promenade has been painstakingly laid by hand in order to create the iconic paving pattern. The generous pedestrian promenade on the south side of the street includes a double row of 240 plane trees, each of which is planted inside a Silva Cell—a modular suspended pavement system that uses soil volume to support large tree growth while still supporting the weight of the pavement above.

Built-in irrigation channels in the pavement steer rainwater to the trees' roots and then the overflow into the sewer system. Altogether, Queens Quay forms a brilliant interplay of programming that allows pedestrians and cyclists to fully enjoy the waterfront area.

Once uninviting, the new world-class Queens Quay links major destinations along the water's edge, creating a public realm that is pedestrian- and cyclist-friendly. The design will create a beautiful and functional lasting legacy for the city.

01 The new bicycle track, provides the missing link in Toronto's Martin Goodman Trail
02 Aerial view showing the relationship between different components of the new street configuration

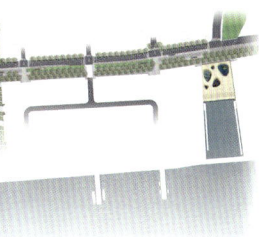

03 The 1.06-mile (1.7-kilometer) Martin Goodman Trail creates a walking and cycling-friendly promenade along the water's edge
04 The new Queens Quay seamlessly merges with the existing iconic wooden Wavedecks
05 Painted maple leaves indicate traffic signals

Elements of the Queens Quay new street profile

04

05

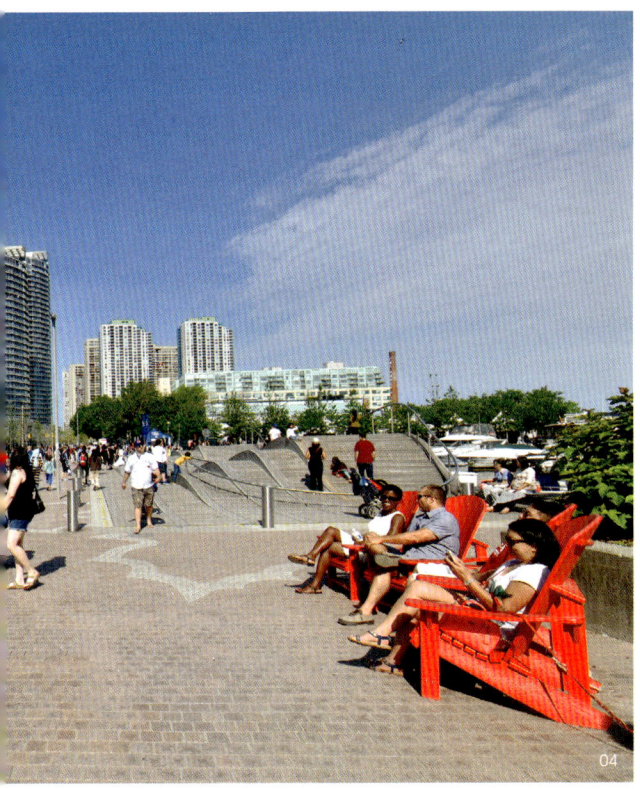

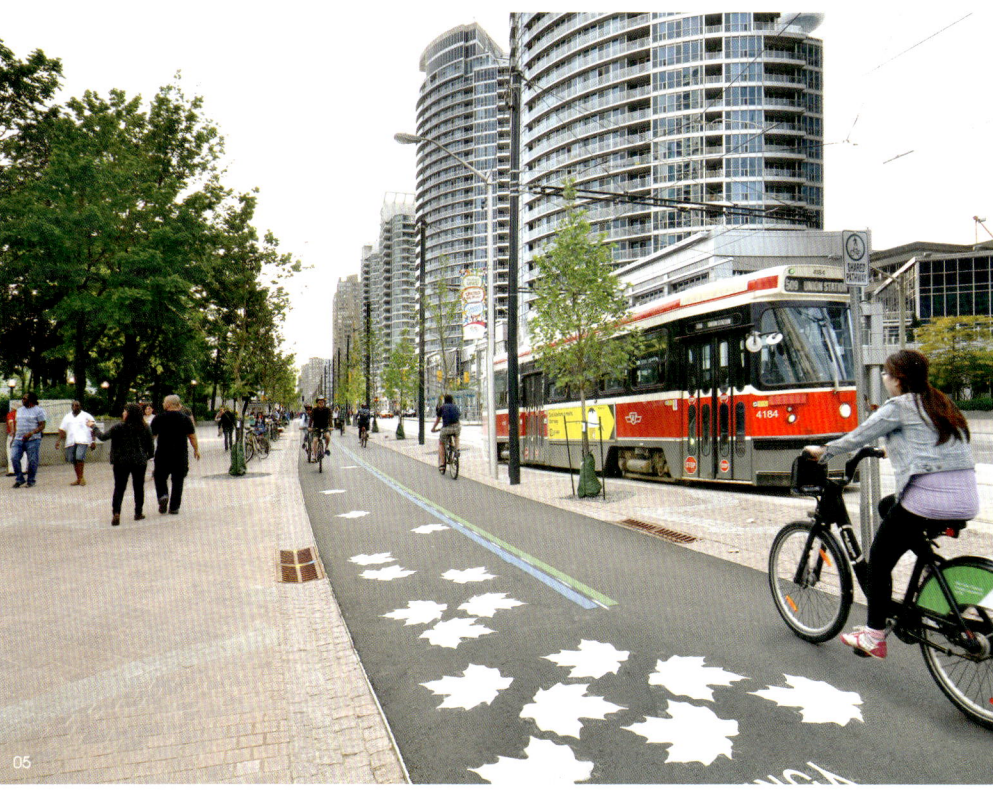

Components of the new Queens Quay

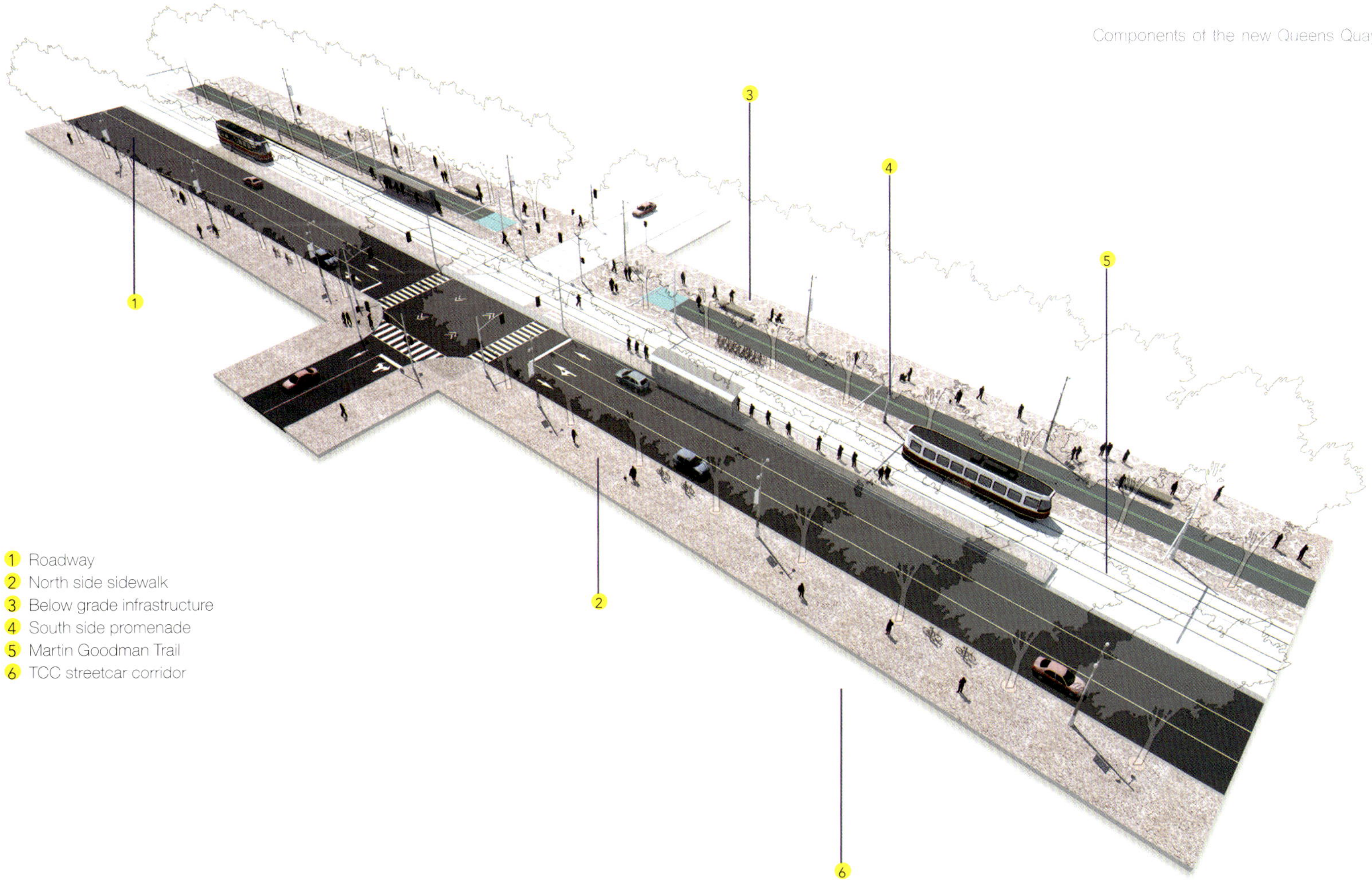
1 Roadway
2 North side sidewalk
3 Below grade infrastructure
4 South side promenade
5 Martin Goodman Trail
6 TCC streetcar corridor

06 Queens Quay at Lower Simcoe looking west
07 The centrally located existing streetcar track has been reconstructed and extended over the full length of the street

08 Toronto's main waterfront street has been transformed into a showpiece for the city
09 The new public realm also features hand-cut granite paving, custom seating and lighting poles
10 This temporary landscape intervention was constructed to allow the public to experience the benefits of the proposed Queens Quay prior to construction

Abercrombie Street upgrade

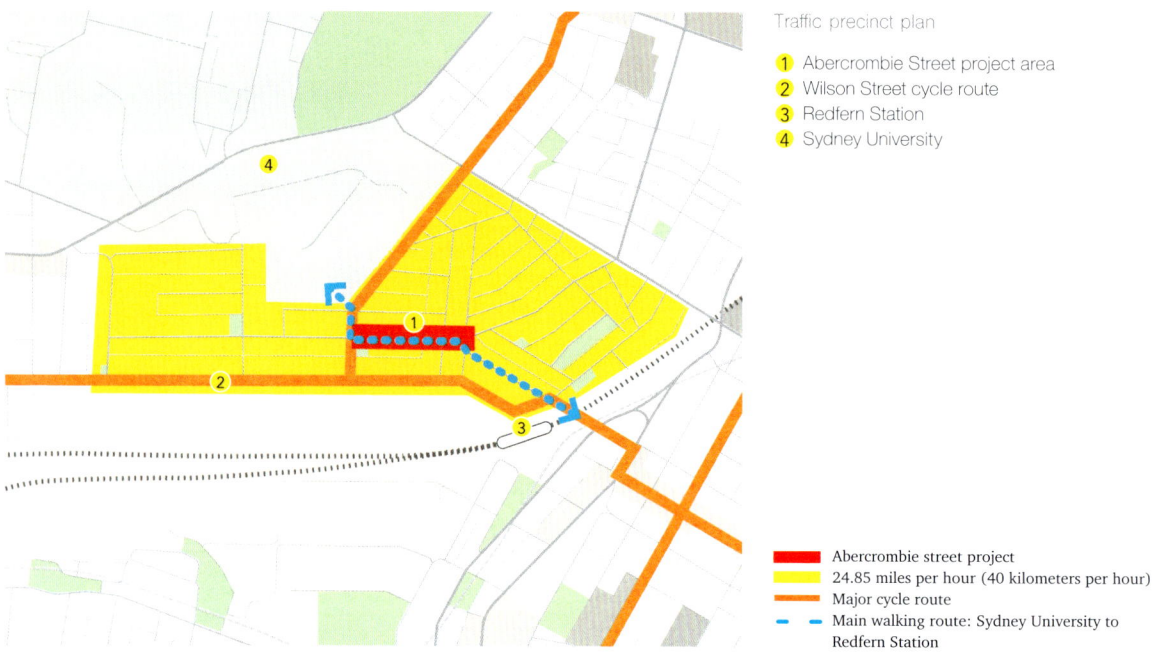

Traffic precinct plan
1. Abercrombie Street project area
2. Wilson Street cycle route
3. Redfern Station
4. Sydney University

- Abercrombie street project
- 24.85 miles per hour (40 kilometers per hour)
- Major cycle route
- Main walking route: Sydney University to Redfern Station

● Location / **Sydney, Australia** ● Area / **1.2 acres (0.5 hectare)** ● Completion / **2015** ● Landscape design / **Group GSA** ● Photographer / **Simon Wood** ● Client / **City of Sydney** ● Budget / **AU$4,300,000**

The revitalization of Abercrombie Street, Darlington, focuses on pedestrians in this busy precinct. This section of Abercrombie Street between Lawson Street and Shepherd Street contains commercial and retail businesses of the Darlington Village, residential properties, and Charles Kernan Reserve. The street also forms part of an important walking route between Redfern Station and Sydney University.

The project is aimed to achieve a streetscape renewal which would re-invigorate the urban environment of the village. The key to the project is the balance between the space provided for significant pedestrian movement and the space for outdoor dining, social life and street greening. The improvements to the Darlington Village include widening the sidewalk, improving underground power lines, establishing new street tree planting and street furniture, and creating a slower traffic environment. The design approach is to value the character of the existing urban streetscape, with emphasis on the retrofit to expand and enrich rather than impose and rebuild. Specific actions are as follows:
- The public domain is expanded within an existing road corridor for outdoor retail and village social life.
- The sidewalk is widened up to 5.6 feet (1.7 meters) along most of the street; underground power lines are improved to increase pedestrian lighting; new street trees grow unimpeded, away from shop awnings, to create a strong avenue which will be a busy pedestrian thoroughfare.
- The greening of the street has been undertaken in a sustainable way; rainwater from roof runoff is diverted through tree and planting areas to provide a sustainable greening of the urban streetscape.

To support pedestrians and village life, additional traffic management measures, including the creation of shared zones and sidewalk continuations along Abercrombie Street, have been taken, which give pedestrians priority when crossing at side streets. Traffic management through the local area has been addressed through the establishment of a slower environment.

Concept plan

1. Footpath upgrade on northern side of the street
2. Existing power lines are placed underground on both sides of the street
3. Shared zones defined by raised pavement at intersections
4. Pavement extension at intersection for pedestrian use and outdoor seating areas
5. New intersection design includes pavement widening on north side, new brick pavement at all corners, and new "smartpoles" providing street lighting, traffic services, and telecommunications
6. Road narrowed to provide wider footpath and more street trees and plantings
7. Raised pavement to enhance pedestrian usability, and stormwater infrastructure improved
8. Shared zones defined by raised pavement at intersections to enhance pedestrian usability

01 New pedestrian paths and lighting

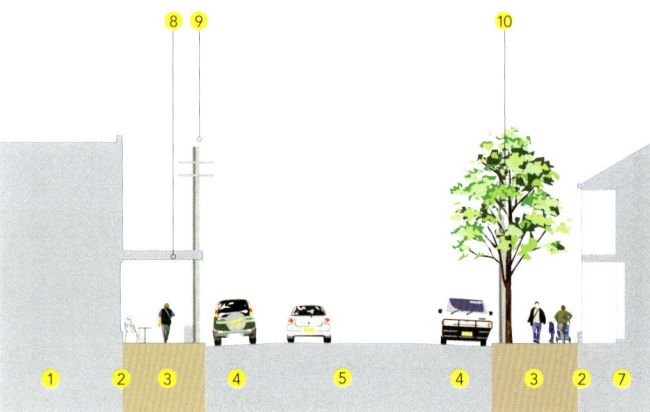
Section A: before

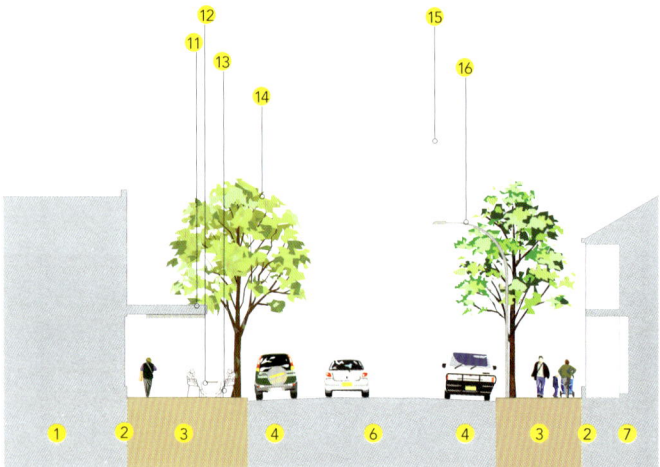
Section B: after

1. Commercial streetfront
2. Boundary
3. Footpath: before, 11.81 feet (3.6 meters); after, 17.39 feet (5.3 meters)
4. Car parking
5. Traffic lanes
6. Narrowed traffic lanes
7. Residences
8. Existing awnings
9. Poles and overhead wires
10. Existing street trees
11. New under-awning lights
12. Café seating areas
13. Widened footpath
14. New street trees
15. Overhead wires removed
16. New street lighting

02–03 Streetscape upgrade to enhance appreciation of built environment
04 Continuous footpath treatment at laneway crossing

New street trees

Rain gardens

Brick pavement

Pedestrian lighting

Seating

05 The public domain is expanded for outdoor retail and village social life
06 Streetscape upgrade beside an adjacent park
07 Widened footpaths with planting and informal seating

1. Existing trees retained within footpath
2. Brick paving and sandstone curb
3. New street along new curb line
4. Garden bed with seating blocks
5. New curb line
6. Existing curb line
7. Pedestrian lightpost
8. Outdoor seating area
9. Awning line

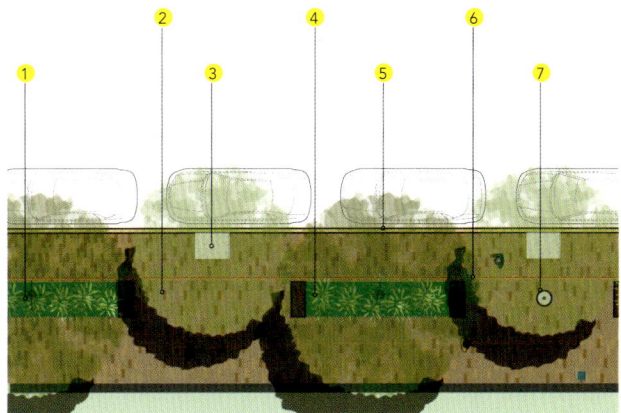

Footpath detail plans: in front of Charles Kernan Reserve

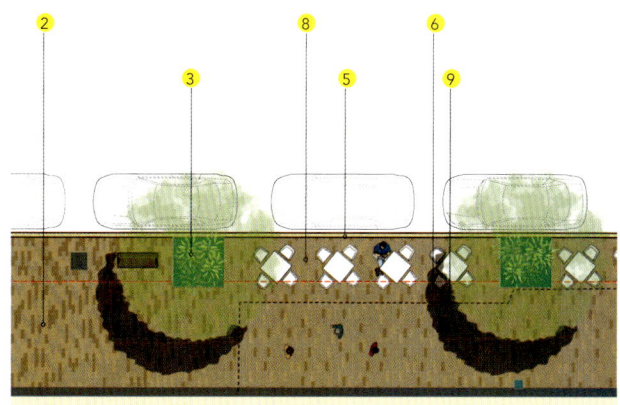

Footpath detail plans: at shop front

Madrid Rio

01 New garden areas around the abutments of the Puente de Segovia that allow access to the area closest to water level
02 Pools and fountains of Puente de Segovia

● Location / **Madrid, Spain** ● Area / **297 acres (120 hectares)** ● Completion / **2011** ● Landscape design / **Burgos & Garrido, Porras & La Casta, Rubio & Ákvarez-Sala, West 8** ● Photographer / **Jeroen Musch, West 8, Municipality Madrid** ● Client / **Municipality of Mardrid** ● Budget / **€410,000,000**

The City of Madrid reclaimed the urban waterfront along the Manzanares River by resurfacing nearly 18.64 miles (30 kilometers) of tunnels and creating 297 acres (120 hectares) of new public space on top of them, which encompasses a total of 47 subprojects, including sports areas, greenswards, plazas, an orchard, an urban beach, children's play areas, bike paths, and eleven new footbridges, which together form an extended green corridor through the existing cityscape, connecting separate green spaces and reclaiming the urban riverfront.

The bicycle network of Madrid Rio is conceived not only for leisure and sports, but also to meet everyday commuting needs. The bike paths total 18.64 miles (30 kilometers), running from north to south and connecting to the city's cycling green belt. At a number of key locations, pedestrian bridges have been built and the existing channels are also utilized to ensure a consistent pathway. The paths are designed to be shared paths for both pedestrians and cyclists. These require patience and understanding to ensure the enjoyment of all users. Priority is given to pedestrians, and cyclists are advised to use caution when cycling. Through improving this pedestrian-friendly connection between the urban districts on both sides of the river, the paths can encourage a new means of moving through the city.

The bridges have the scale of park elements. The Cascara Bridges are designed as massive concrete domes with a rough texture. Their slim steel decks are suspended by more than one hundred cables that resemble the mouth of a baleen whale. The fine detailing becomes visible when visitors enter the bridge: the ceilings are embellished with a mosaic by the Spanish artist Daniel Canogar, and feature custom lighting that brings the artwork and the deck alive at night.

Madrid Rio is part of the larger vision for the city that embraces culture, education and well-being for its future restructuring. It transforms an area that was historically the "back door" of Madrid into a park that bridges neighboring communities, and functions as a platform for various activities and healthier lifestyles. The bike and circulation network of the park is a significant part of the whole Madrid Rio project.

Cycling routes of Madrid Rio

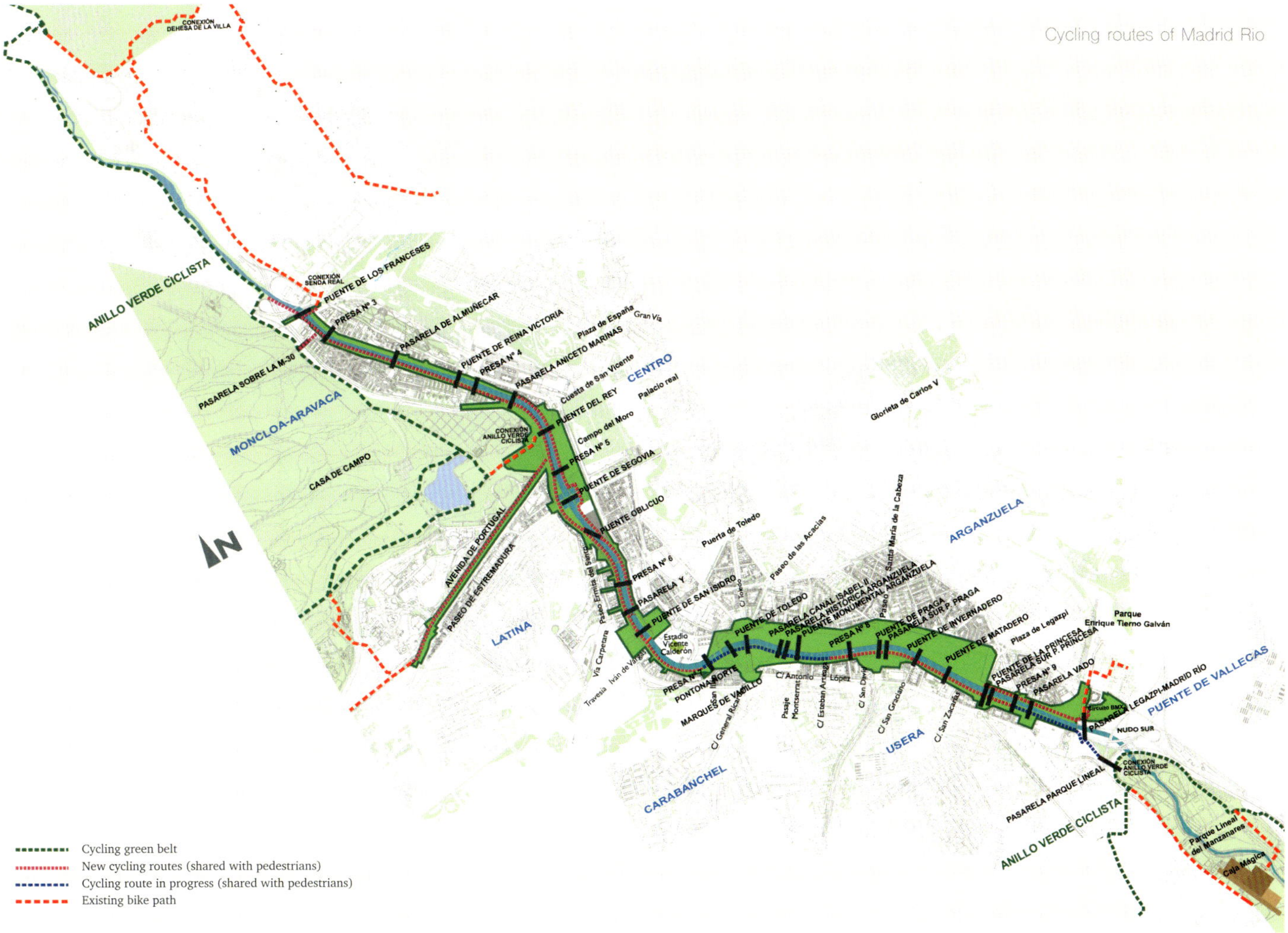

- - - - - - Cycling green belt
- - - - - - New cycling routes (shared with pedestrians)
- - - - - - Cycling route in progress (shared with pedestrians)
- - - - - - Existing bike path

03

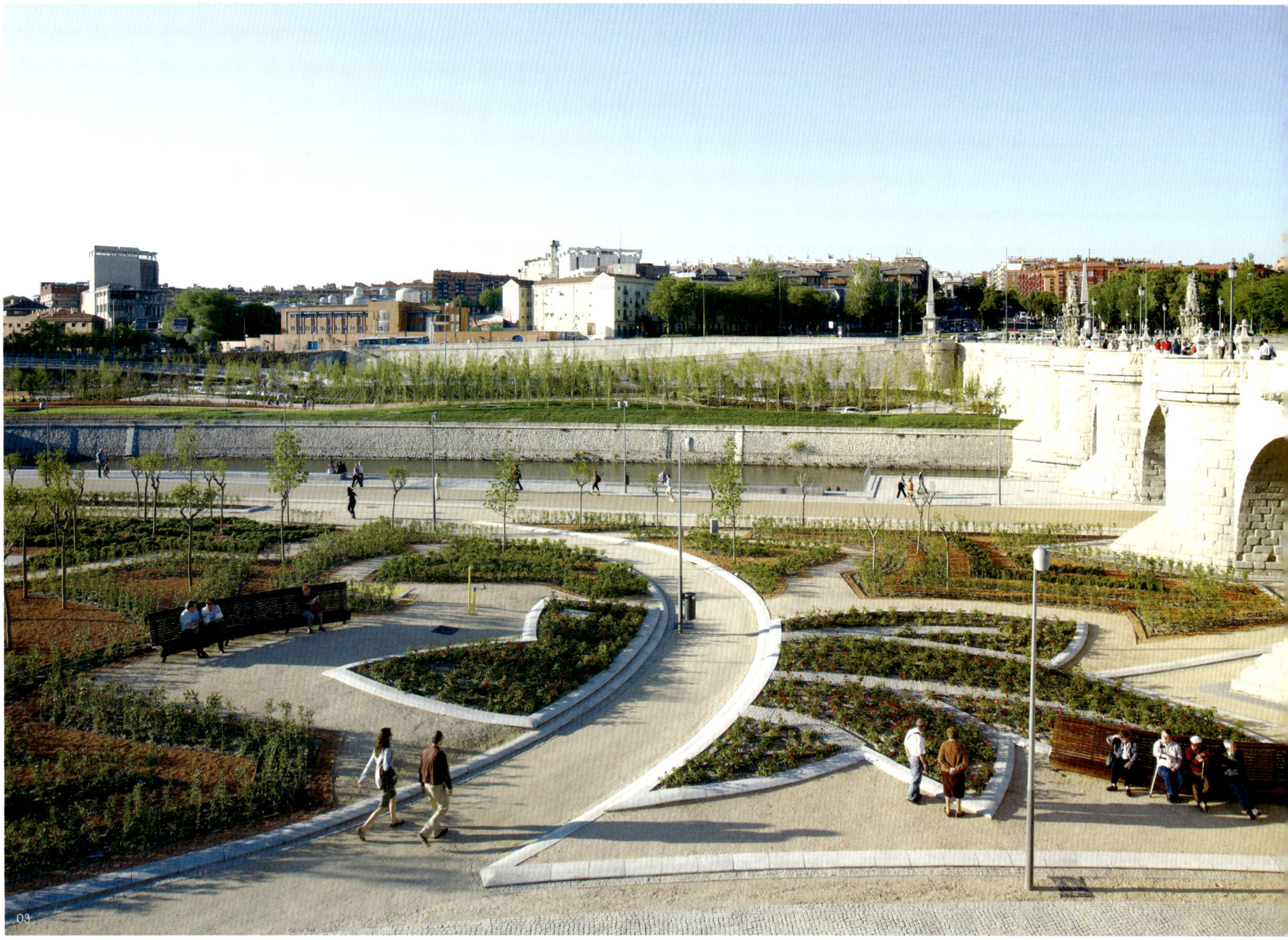

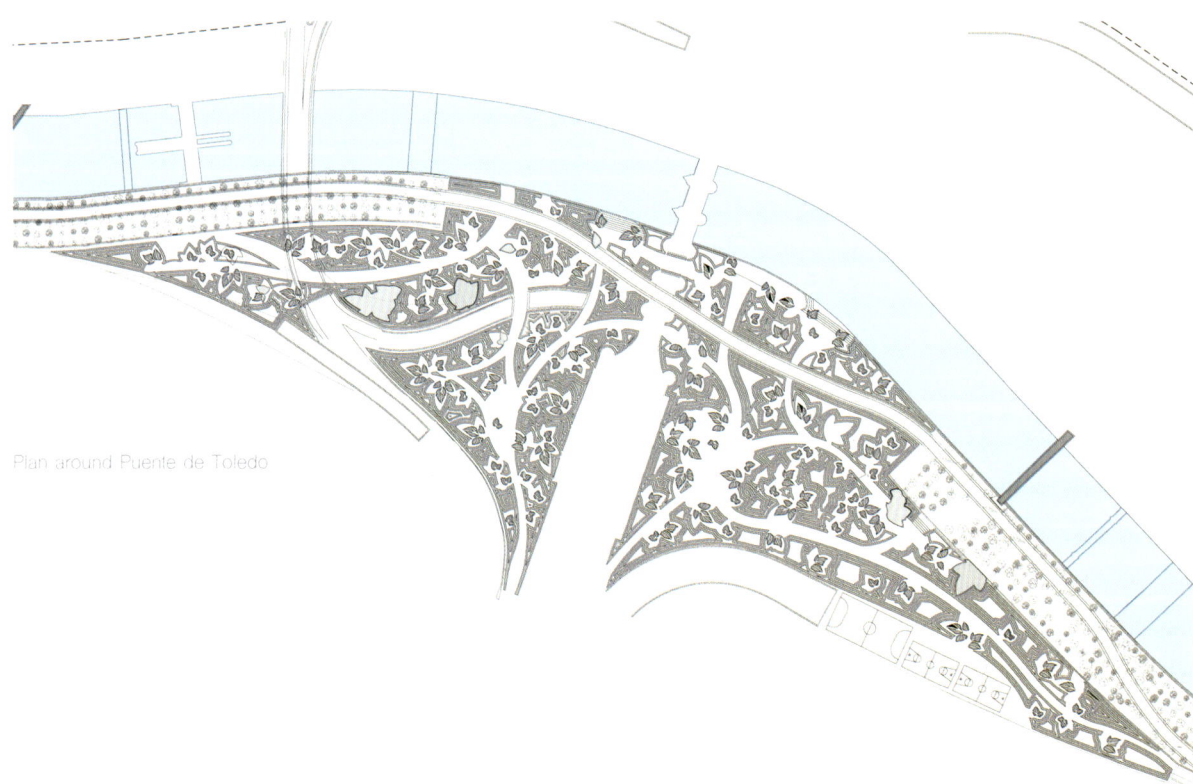

Plan around Puente de Toledo

03 The gardens below the Puente de Toledo
04 The riverside promenade offers pedestrians the unprecedented experience of walking beneath the bridge's stone arches
05 Along the Salón de Pinos, a collection of small playgrounds form a coherent and articulated group in the linear landscape

06 Platforma Puerta del Rey
07 The ergonomic shape of custom-designed planters allows them to be used as benches or seats
08 The Salón de Pinos has become a dynamic access axis for walking and cycling
09 The cherry motif is outlined in the small basalt and limestone mosaic paving of the Avenida de Portugal

Section of the Huerta de la Partida

1 Fruit trees (plums and pears)
2 Edge/seating element at stream edge
3 Stream
4 Edge/seating element
5 Fruit trees (apples)
6 Huerta de la Partida
7 Riverside
8 Path

(10) The bridge features a double-curved reinforced concrete sheet from which the metal deck structure is suspended
(11) Children's play area under the viaduct located to the north of the Puente de Toledo
(12) Cascara Bridge with mosaic ceiling by Daniel Canogar
(13) Avenida de Portugal

10

11

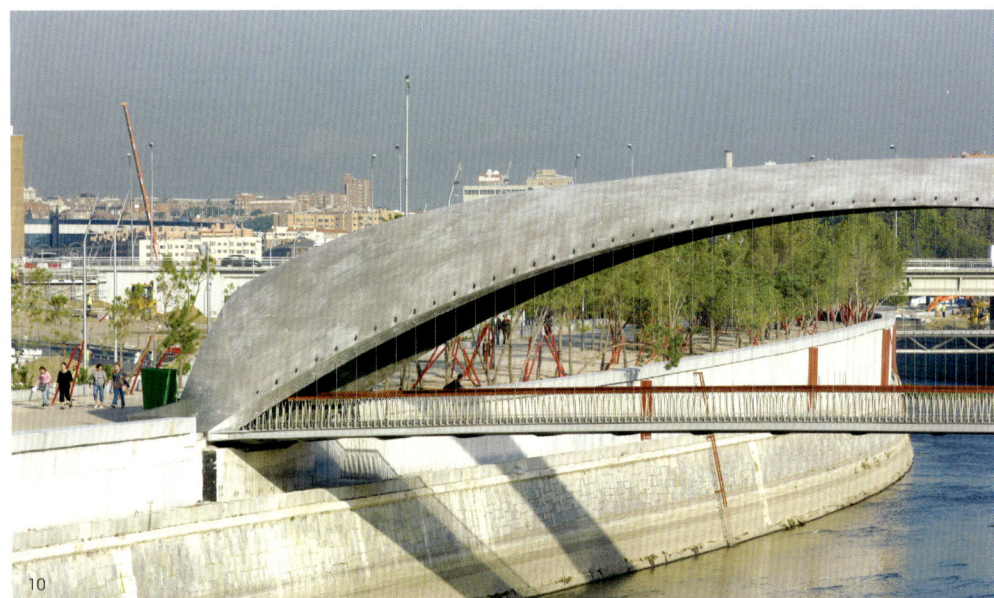

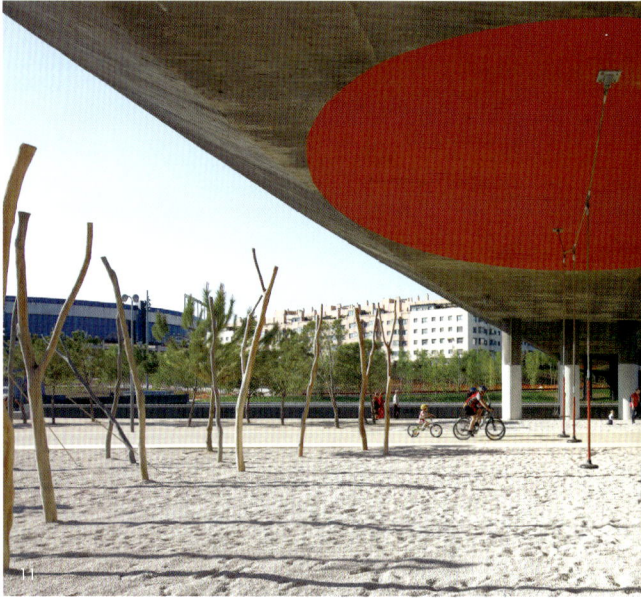

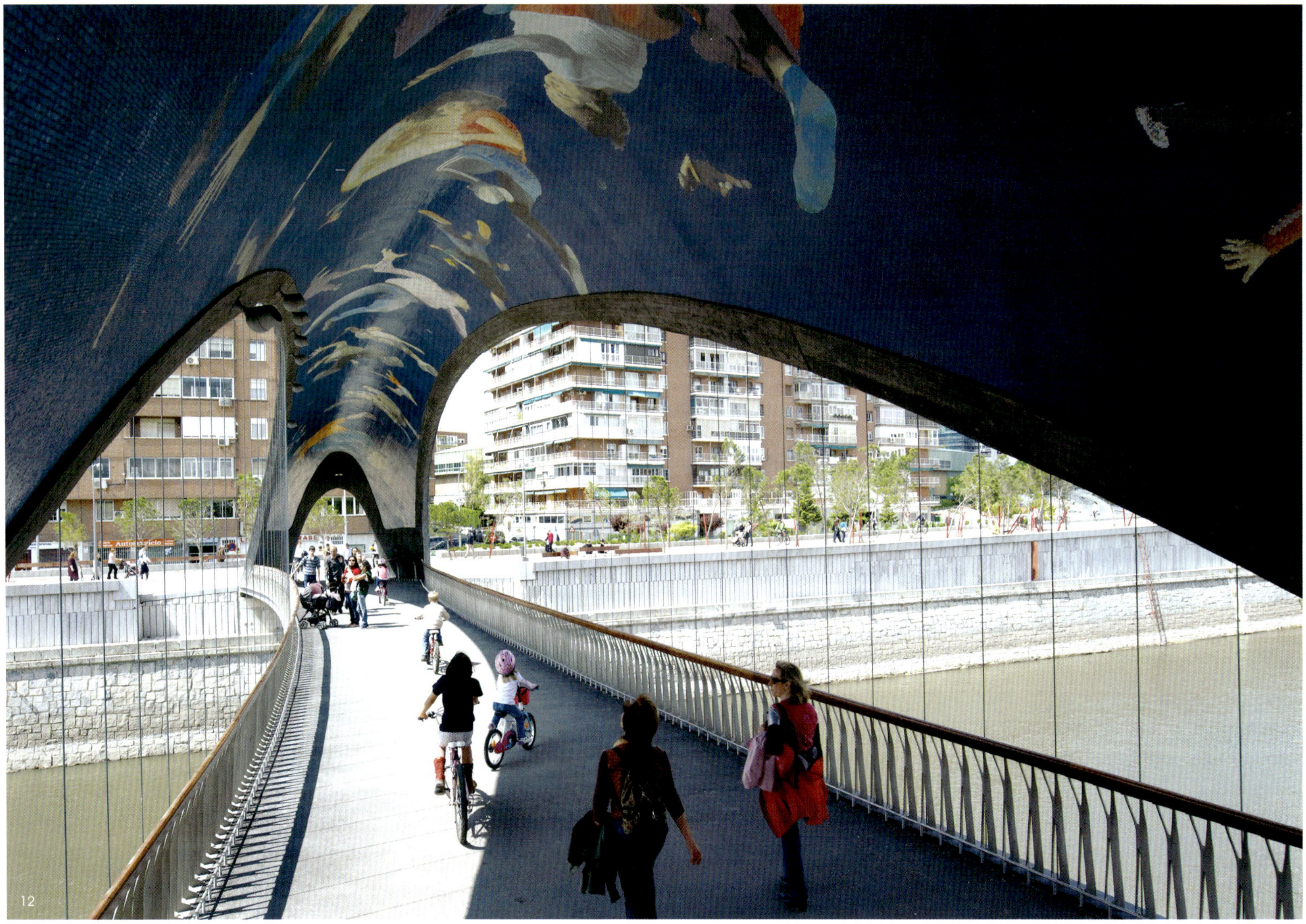
12

General cross-section of Salón de Pinos

1. Sidewalk
2. Parking
3. Avenida del Manzanares
4. Bicyle pathway
5. Salón de Pinos

Governors Island,
Phase 1—Park and Public Space

Master plan

1. The Play Lawn
2. Hammock Grove
3. Liggett Terrace
4. Historic North District
5. Soissons Landing
6. Yankee Landing

● Location / New York, United States ● Area / 30 acres (12 hectares) ● Completion / 2014 ● Landscape design / West 8, Mathews Nielsen Landscape Architects ● Photographers / Timothy Schenck, Kreg Holt, Stacey Shand Fletcher, Iwan Baan, Jim Navarro, West 8 ● Client / The Trust for Governors Island ● Budget / US$70,000,000

The Governors Island, Phase 1—Park and Public Space project has dramatically transformed this abandoned military base and forgotten island into a dynamic public space. Inaccessible for two centuries, Governors Island is now New York's summer playground. Phase 1 of the project—encompassing 30 acres (12 hectares) of new park—includes Liggett Terrace, Hammock Grove, and the Play Lawns, all bustling with activity.

Visitors can cycle on the island's meandering paths, swing with friends on clusters of custom hammocks, stroll on the promenade at the water's edge, or play baseball with the Statue of Liberty as a backdrop. The Governors Island Park and Public Spaces are a car-free oasis. The island features a clear, organized mobility circuit, ensuring the safety of all users. West 8's design encompasses a network of paths and roads with a hierarchy of uses, which serve pedestrians, cyclists, and service vehicles, but gives priority to pedestrians and cyclists.

Paths, roads, and intersections are designed to minimize conflict between all road users. Cycing and walking paths are not accessible by park-serving vehicles except in the case of emergency. Cycling is integral to the island experience. It is the fastest way to explore the 5 miles (8 kilometers) of paths (including the Great Promenade) that criss-cross the Island. Paths are gently curved and spaciously sized to accommodate easy, leisurely riding and appeal to first-time riders and families riding together. Markings, signage and path widths separate bicycle routes from pedestrian routes, and 3.2 miles (5.2 kilometers) of precast concrete curb edges also offer seating for those who wish to stop along the way.

This formerly abandoned military base and forgotten windblown island has been transformed into a novel and affordable destination. The people-centered design embraces active and environmental design philosophies, offering a rich array of active and passive recreation choices and a constantly evolving program supported by robust walking and cycling infrastructure, which has seen Governors Island become a destination for the people of New York, the "dot" on Manhattan's exclamation point, and the centrepiece of the city's efforts to revitalize its waterfront.

01 Governors Island is a 170.50-acre (69-hectare) island in Upper New York Bay, approximately 800 yards (73.15 meters) from the southern tip of Manhattan
02 Ligget Terrace is punctuated with seasonal plantings, water features, children's play area, and artfully crafted mosaic paving

Circulation plan

Circulation types
- Pedestrian only
- Pedestrian and bicycle
- Pedestrian, bicycle and vehicle
- Public access corridors in development zones

03 The wide pathways of sinuous curves and gently sloping topography encourage visitors to explore the island on foot and by bike
04 Framed by historical architecture Liggett Terrace, a 6-acre (2.43-hectares) outdoor "room", is a lively entry into the southern island
05 Small high points in the landscape accentuate visitors' sense of discovery and connection with the wind and the harbor
06 This project protects the island as an oasis of car-free cycling, with adults and children's cruiser bikes available for free every weekday from 10 am to noon

06

06

The park and public spaces are an integrated system of layered components

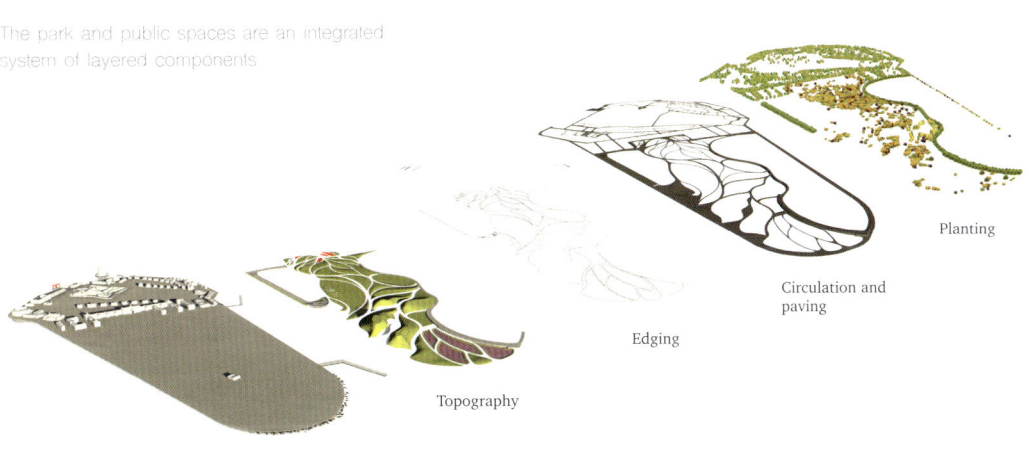

Existing island Topography Edging Circulation and paving Planting

07 The logistics of ferry queuing, servicing, and disembarkation, such as that at Soissons Landing, have been informed by empirical research and direct observation

08 Seasonal plantings add color and interest to the new park and public spaces, supporting new habitats

09–10 More than 3 miles (5 kilometers) of embossed precast edging acts like eyeliner to accentuate the composition of views, undulating pathways, and forest

New Park for the University Quarter in Essen

Master plan

● Location / Essen, Germany ● Area / 12.60 acres (5.1 hectares) ● Completion / 2014 ● Landscape design / scape Landschaftsarchitekten GmbH ● Photographers / Matthias Funk, Rainer Sachse, Hans Blossey, WAZ FotoPool Walter Buchholz ● Client / Grün und Gruga Essen, Entwicklungsgesellschaft Universitätsviertel Essen ● Budget / €6,400,000

The University Quarter is developed on a former railway site in the inner city of Essen, which had been wasteland for almost 30 years. By revitalizing this 32.12-acre (13-hectare) plot, the University Quarter—Essen's Green Center developed into an urban quarter for residential and commercial use. Following an independent design language that is derived from the typology and history of the site, the park is unique. But the focus here is on the design of the promenade in the park.

The surrounding residential and commercial buildings form an urban figure that gives the park a very long, narrow, and radial shape. One of the most important design statements is about the arrangement of attractions and paths based on the consideration of the unusually shaped park area, the request for water basins and varied recreational activities, as well as the expected high usage of the inner-city park. To reduce potential conflicts between residents and park users, secondary paths along the park edges strengthen the semi-public character.

The main promenade is located in the spacious center of the park along with recreation, sport, and play facilities. This design not only creates a central, open, public promenade between the northern water park and the southern lawn park, but also protects the residents from noise. To reinforce the spatial attraction of the promenade, it is not set out on one axis, but branches out several times. The Promenade Places—attractive destination points—are located in those branches. The strip of water basins orientates itself with the shallow, accessible water side toward the promenade, and therefore reinforces the public nature of the scheme. The promenade is linked to the surrounding plots by a multitude of cross paths through the lawn areas and footbridges over the water basins. An open, multi-functional area for events with a large flight of steps is located at the crossing point of the promenade with the sidewalk connecting the university and the city center.

① Rounded edges determines the entire urban design by referring to the local foundry industry
② Consistent design language of unique benches harmonizes with the urban design

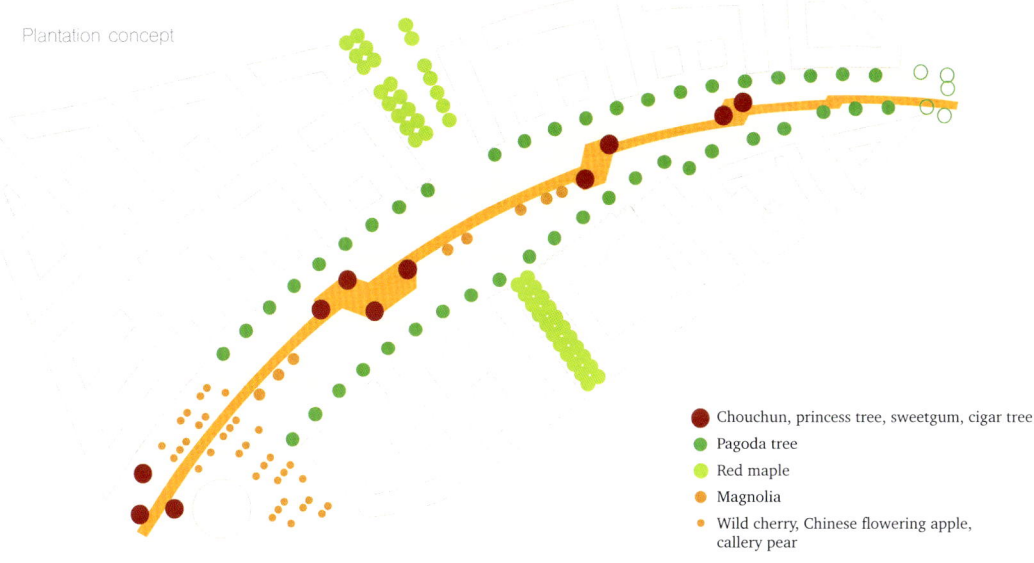

Plantation concept

- Chouchun, princess tree, sweetgum, cigar tree
- Pagoda tree
- Red maple
- Magnolia
- Wild cherry, Chinese flowering apple, callery pear

The design and choice of materials is deliberately technical. As the entire park is constructed in the shape of a curved grid, all visible edges and radii are defined in a very simple, mathematical way. The impression of the whole park as one homogeneous iron element is achieved by lifting the path areas (mastic asphalt) with sloping granite boards off the base. This design feature is repeated at the Promenade Places which are, in turn, lifted off the path areas.

The promenade is accompanied by some tree magnolias. The Promenade Places along the promenade are emphasized by large trees such as empress trees, American sweet gum, and Indian bean trees. Attractive perennials and bulbs accompany the central promenade. The central promenade is brightly illuminated, while the paths to the sides are a bit darker, but still sufficiently lit for traffic. The Promenade Places are accentuated by directed lights on different angles. They set a stage-like character, which also makes them attractive places at night.

(03) Footbridge over the water basin
(04) View along the promenade along the waterside
(05) View along the promenade

169

06 Wide paths connect the surroundings to the site
07 Promenade along the waterside
08 Water basin attracts locals and visitors
09 Main promenade is linked to the surrounding plots by footbridges over the water basins
10 Japanese pagoda trees are planted in an east–west direction

Section

⑪–⑭ Material details
⑮ Perennials and ornamental grasses bring warm colors into the park during the autumn
⑯ Next to the water basin, softly elevated parts are equipped with benches, offering opportunity for social gatherings

Torreblanca Aromatic Park in Torrevieja

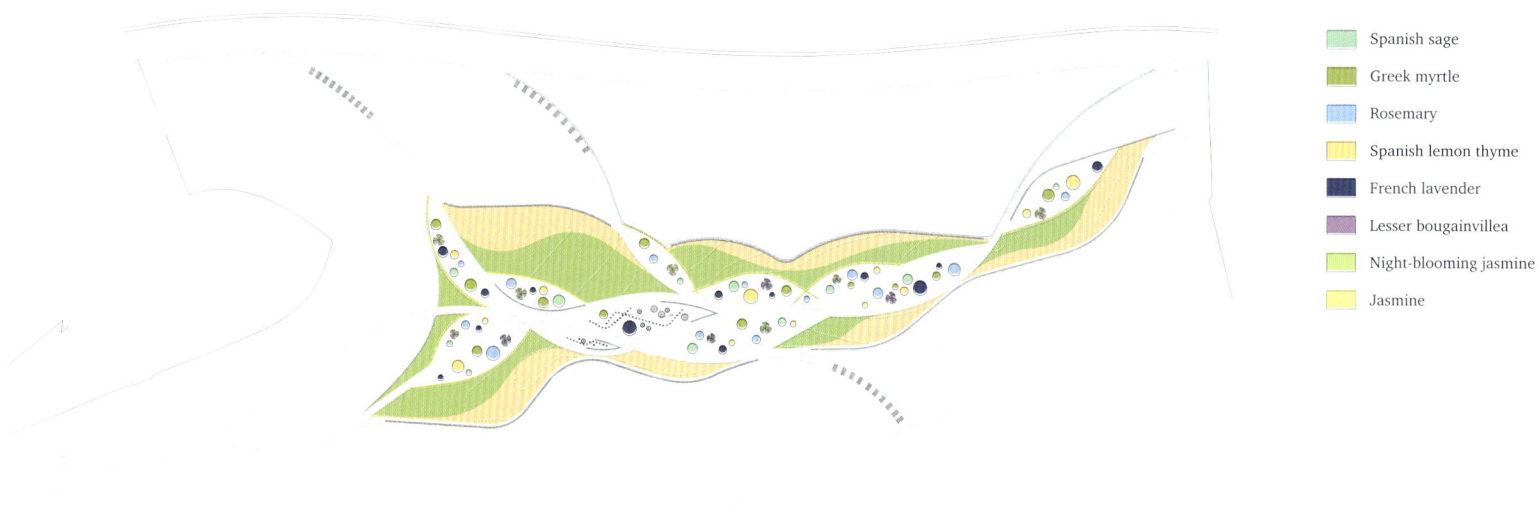

- Spanish sage
- Greek myrtle
- Rosemary
- Spanish lemon thyme
- French lavender
- Lesser bougainvillea
- Night-blooming jasmine
- Jasmine

Treatment to vegetal parterres

● Location / **Torrevieja, Alicante, Spain** ● Area / **18.61 acres (7.53 hectares)** ● Completion / **2010** ● Landscape design / **Estudio Carme Pinós** ● Photographer / **David Cabrera** ● Client / **Ayuntamiento de Torrevieja** ● Budget / **€1,664,425**

This marginal lot is left over from the uncontrolled growth of Torrevieja. The strong contour change in both sides of the lot makes the construction very complicated. The project consists of replanting the slopes with pine trees, turning them into a forest where small paths go down to the lower level. This level is where the paths entwine with aromatic planting beds, always in circles. Sometimes, these circles are large, and metal trelliswork supports vines.

The first major step was to consolidate the surrounding slopes of the plot. The center of the valley was flattened to accommodate the intended beds and the trail paths were plotted out.

Given the slopes of the hillsides, there had to be a drainage system to limit the effects of erosion from runoff. Trenches perpendicular to the slope, which are placed around the perimeter of the plot, aid this effort. Excess runoff can escape through gravel pits about 6.56 feet (2 meters) deep, located on the slopes and scattered around the perimeter. Any other unabsorbed water is collected by PVC subcollectors, which have a diameter of 11.81 inches (30 centimeters). These feed into a central concrete pipe measuring 19.69 inches (50 centimeters) in diameter. This central collector is located in the center of the park. After excess water enters the collector, it passes through a grease trap and then into a longitudinal tank buried in the bottom of the park, before being evacuated to the spillway north of the plot.

A drip-type irrigation system takes advantage of the steepness of the slopes to supply the pine forests with water. A 1.97-inch-diameter (50-millimeter-diameter) tube traveling around the perimeter of the plot is connected to a number of polyethylene tubes placed in the same direction as the slope. The second part of this system involves sprinklers positioned to irrigate the flattest part of the plot: the center of the valley where the different beds are located.

There are two main circuits positioned along the main roads. All of the power to the park lights is supplied by these circuits. A single control panel (placed in the center of the park and hidden off to the side) controls the two circuits. Other mast-style lights made of several steel columns are placed along the roads and at the entrances. The other "simes" style lights are embedded in the foundation near the banks placed at the edges of flower beds.

Layer organization

Paths	Stone areas
Banks and clay	Flower beds and water
Lawn	Whole area

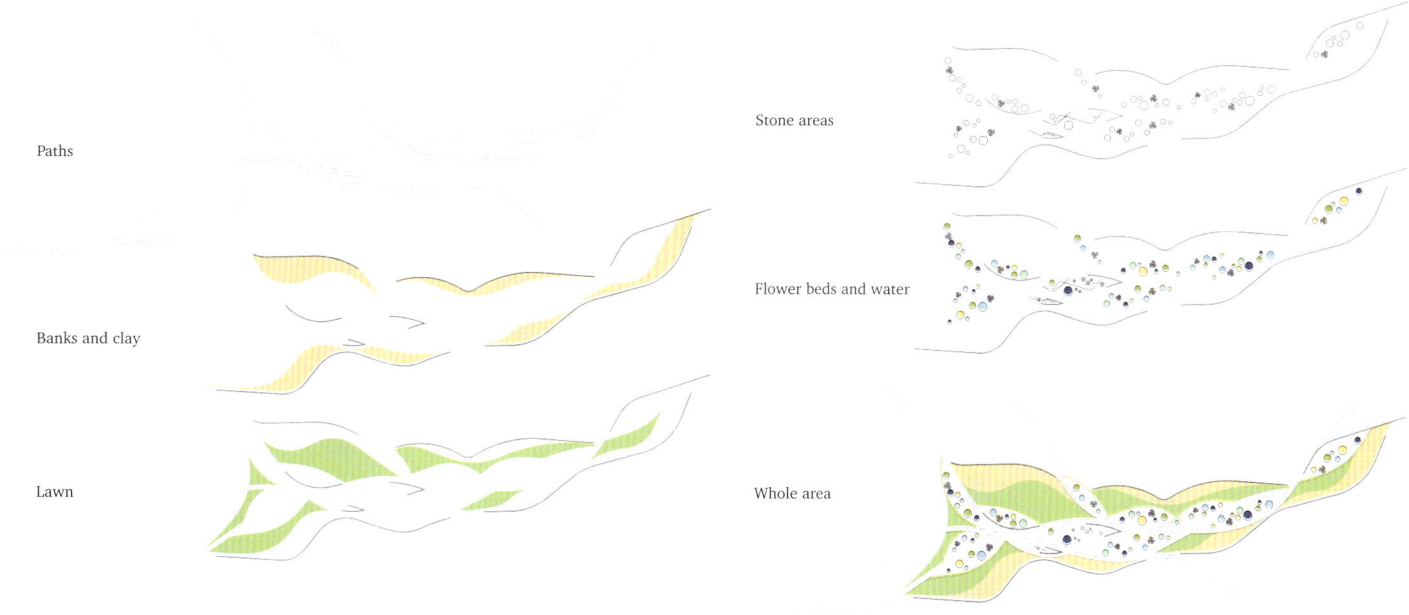

① View from above in northeast direction

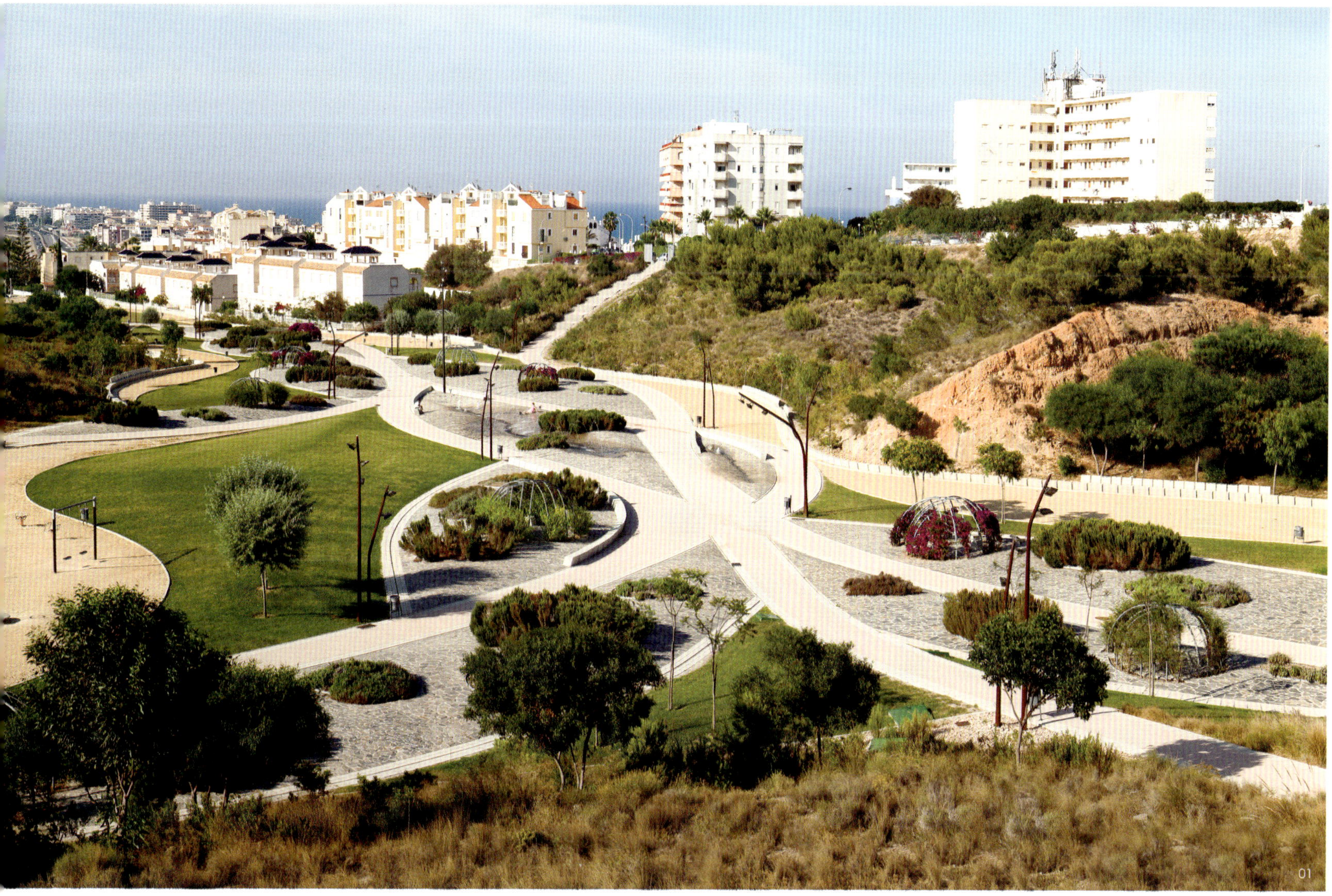

The designer wanted to provide two types of paths. One allows direct, highway-like travel through the park, while the other navigates the outline of the flower beds. The main roads are marked out with precast concrete pieces measuring 7.87×15.75×1.97 inches (20×40×5 centimeters) with a crushed gravel center. Where slopes are greater than 8 percent, the road becomes a ramp-staircase consisting of precast concrete pieces measuring 7.87×15.75×7×1.97 inches (20×40×18×5 centimeters). The secondary paths are simply crushed gravel roads, while the sidewalk along the perimeter of the plot of the park is made of 1.97-inch-thick (5-centimeter-thick) colored concrete slab on a 3.94-feet-thick (1.2-meter-thick) concrete base. The edges of this sidewalk are composed of precast concrete pieces of 7.87×15.75×7.87 inches (20×40×20 centimeters). The entrances to the park are defined by a widening of the sidewalk overlapping the main path.

Benches are located throughout the park. They are found in the boundary areas between the slopes and the flattest area of the beds. They are used not only to help position, but also to hide the aforementioned gravel ditches, which collect runoff from the slopes. All the park benches are backed with prefabricated concrete set in the same foundation by steel flat plates. The archway-style climbing plants are arranged in the parterres in a way to allow for their eventual size. These parterres consist of spiraling metal arches. Upon this structure are round spiral devices, along which grow some vines.

02 Playground, paths, and aromatic plants parterres
03 Benches and parterres areas

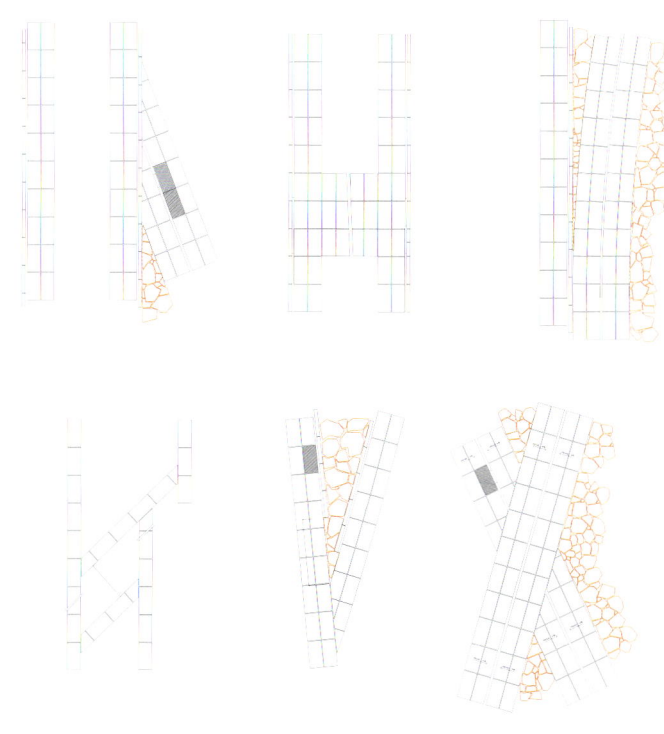

Crossed-path details

04 Paths interweave to form stone areas where the parterres with aromatic plants are placed
05 Night view of the valley and park
06-07 Paths between parterres
08 Structure of metal arches which provide a quiet place to rest

The Erbe Danzanti Park

The paving scheme sketches: railway segments are embedded in the porphyry pavement

● Location / **Paratico, Italy** ● Area / **2.35 acres (0.95 hectares)** ● Completion / **2010** ● Landscape design / **Cristina Mazzucchelli Green Design** ● Photographer / **Cristina Mazzucchelli** ● Client / **Comune Di Paratico** ● Budget / **€950,000**

A lacustrian park is born in Paratico, on Lake Iseo's shores, Northern Italy. It was an unused terrain previously belonging to Ferroviedello Stato—the Italian National Railways. The project created a public recreational area while highlighting the beauty of the environment and restoring traces of the location's past. The relics of the railways have been both preserved and highlighted in the design of the walking paths in the park, which serve as a "keynote" connecting several green areas within the park spiritually and physically, while each of these areas focuses on a different theme, striving towards interpretation of the location's many souls.

The park develops, uninterrupted, along the lake's shores with a sequence of "rooms" or sections. Two lengthwise walking paths cross the park. The first path is a straight promenade, extending along the shore, for people to stroll through the park while enjoying the lake view. The second one cuts right through the park in a diagonal way, as it purposefully retraces the old railway lines. Following the second path, people can find hedges in bloom and large metal tanks brimming with graceful plants placed in a diagonal section, leading the eye towards the lacustrian environment; the room where the gravel draws a pattern of waves, reminding one of the waves rising from the lake's rippled waters; the room dominated by a large wooden arbor, covered in American grapevine embracing scented roses, hinting at the farming tradition of those Franciacorta hills, cultivated for their grape crop; and the final landmark, a room containing numerous seats, stemming from an oval-shaped design, and two rectangular stone tanks with white water lilies, thus recalling concisely all the different sceneries the lake offers along its shores.

The pedestrian trail is paved with local stone settled with an irregular scheme that follows the "Roman way" in order to create rhythm. The trial is characterized by the presence of the old train tracks embedded between the stones, which accompany the entire length of the path. Combinations of colors, shapes, and patterns create a feeling of great spontaneity to an overall view of the park, hinting at harmony, gracefulness, and movement. This is how the park got its name. This project serves as a good example of how to integrate landscape, history, and leisure.

Master plan | The vibrant flowers room | The waves room | The grape tunnel room | The waterlily room | The restaurant room

1. Charcoal shed converted to a restaurant
2. The flowering pear-tree wood
3. Raised flower beds
4. The rounded resting place
5. The tunnel covered by roses
6. Parking
7. Wooden pergola
8. The oval stopping place
9. The waterlily pools
10. Restaurant
11. The restored jetties
12. Restored stone piers

▶ Main entrance
▷ Secondary entrance
▬ Lawn
▬ Perennial beds

01　The two lengthwise routes that cross the park purposefully retrace the old railway lines; the central one passes through a sequence of different "rooms," each one uniquely designed
02　The gravel in this park room draws a pattern of waves, alongside spacious fields, which soften the effect of the stones

03 The main entrance of the park, where the two lengthwise paths start: the first follows the shore, whereas the second one cuts right through the park
04 The path that follows the shore is delimited on one side by an uninterrupted sequence ot Stipa tenuissima and Verbena bonariensis, which keep people a safe distance from the shore edge

05 The central path passes near an iron tunnel covered by clematis and roses and then into a square wooden arbour, covered in American grapevine and scented roses

06 The view to the lake is enriched by a rectangular stone tank with white water lilies and fish. The restored stone piers allow bathers to enjoy the lake

- - - Main lengthwise path, retracing the old railway lines
- - - Secondary cross path
- - - Path along the parking areas

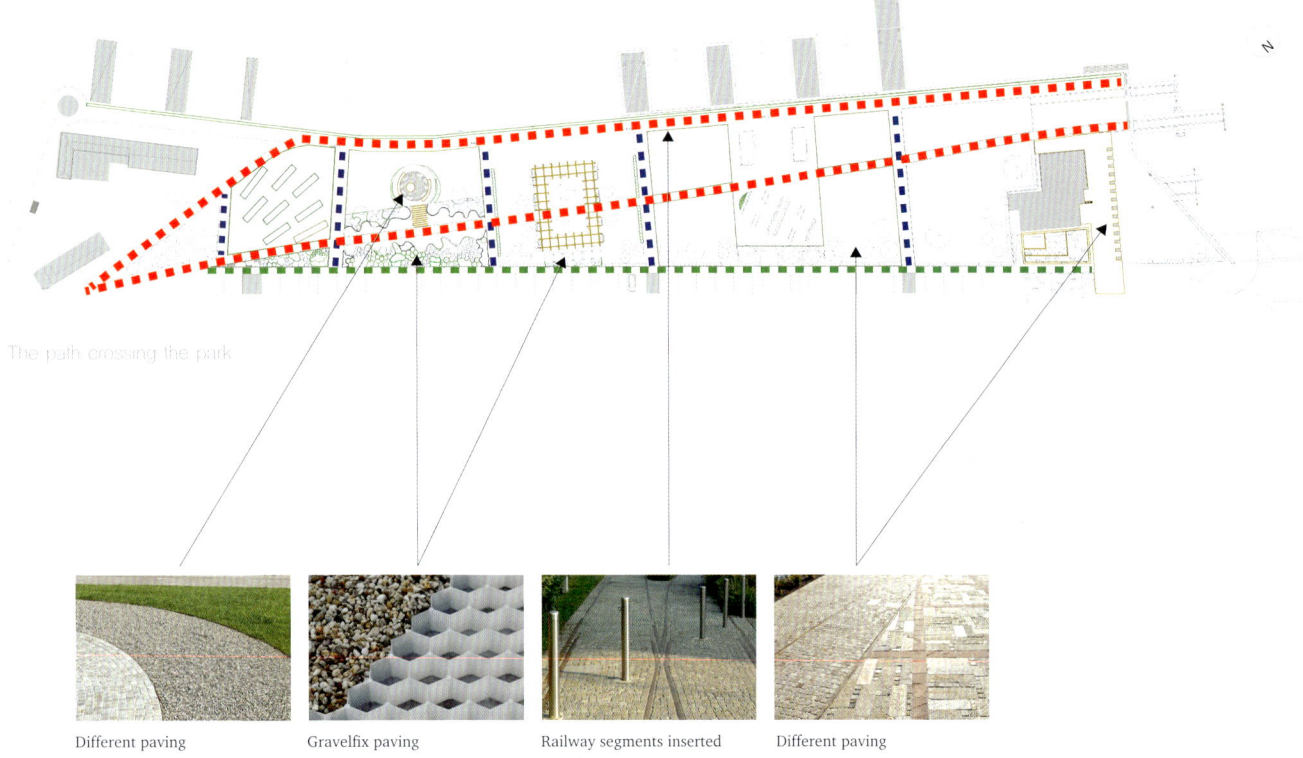

The path crossing the park

| Different paving | Gravelfix paving | Railway segments inserted into the paving | Different paving |

07 The two independent paths converge on a square where the restored jetties create the focal point, leading the eyes towards the lake and the surrounding mountains

08 – 12 Details of the waves room: the gravel has been stabilized using a honeycomb plastic support to create drainable, firm and long-lasting pavings (ENKADRAIN); a circular square, furnished with wide wooden seats, is paved with porphyry stones following a concentric pattern; the gravel is separated from the stone surfaces by an iron edge

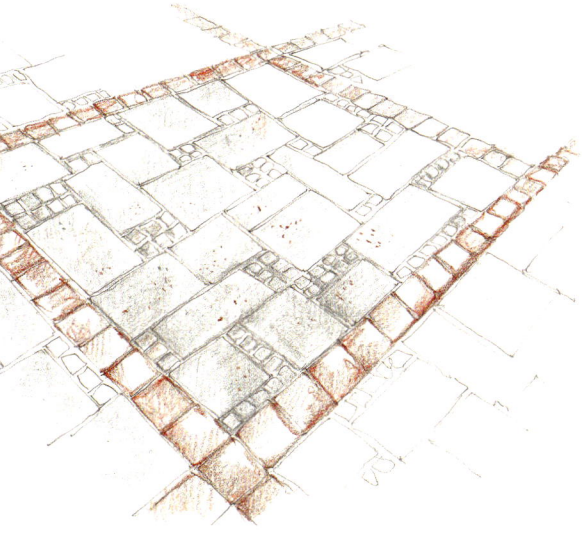

Typical square scheme sketch (6.56 by 6.56 feet [2 by 2 meters]), composed of red and gray porphyry and "Luserna" stones of different shapes and sizes, laid according to the "Roman style"

The Porto-Lerone Cycle Lane

The cycle lane plan

● Location / Arezano, Italy ● Area / 1.48 acres (0.6 hectares) ● Completion / 2011 ● Landscape design / CANAPÉ Cantieriaperti—Architecture and Urban Design ● Photographer / Sergio Fortini ● Client / Arenzano Public Administration ● Budget / €300,000

This project focuses on the realization of a walking/cycling lane, from the Port of Arenzano to the sports and playground area near the mouth of the Lerone River, running for 951.33 feet (290 meters) along the layout of an old railway. Actually, the disused railway is changing into an urban space that encourages sustainable mobility.

The project's strategy was to consider the landscape as the most important influence on the characterization of the path. The most awesome element of the project area is the sea, a permanent scenery all along the bike path. The proposal was to underline the strengths of the site through the individuation of different kinds of spaces for social activities, free time, and recreation, but not to consider the path as a simple line between two different points. The bike path spreads out in new urban spaces for the users, which are created with materials in light of the landscape patterns: the solid ones, Corten steel and wood, and the dynamic ones, gravel and sand, to invoke the nearby seaside. The selection of pavement materials for the bike path has taken the continuity between the artificial elements and the landscape into consideration. It adopted artificial stone, which is neutral and continuous, as a natural sand track near the sea.

The project also includes some green areas composed of Mediterranean plants. The design of the vegetation satisfies different functions: trees (Tamarix Gallica) are planted near the rest areas, and shrubs near the artificial hills. In place of an old car park, a possibly hidden location for the users of the lane, an extended common bench and a set of Tamarix Gallica mark a public area near the restaurant. The vegetation generates a system of green urban places shaping the landscape for the path. The artificial hills are created by recycling materials from excavation, according to a sustainable urban design. The maintenance of the materials and the irrigation system also satisfies the requirements of sustainability, recycling, and economy.

Technological section 1

1. Steel profile
2. Gravel
3. Wooden path
4. Concrete containment
5. Sand
6. Artificial stone path
7. Drainage system
8. Wooden bech

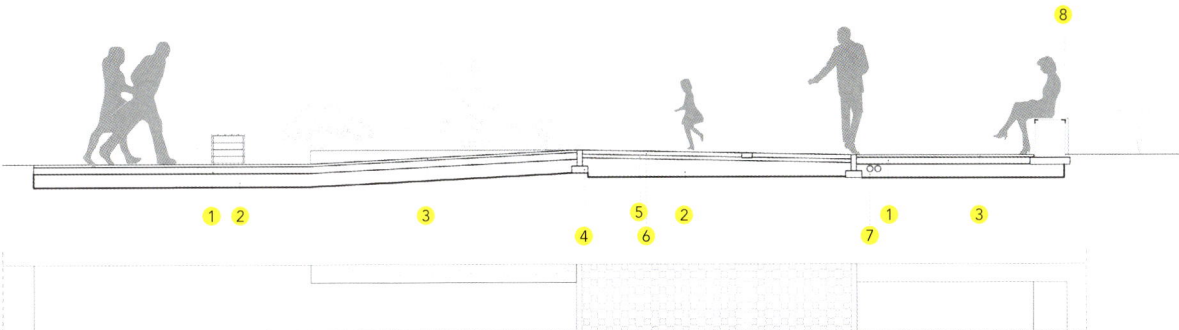

① Instead of demolishing the existing fence, an iron structure is used to cover it up, which is holding up a line of climbers, forming a continuous green skin
② The material of the path (artificial stone) is neutral and continuous, like a natural sand track near the sea

Technological section 2

1. Wooden bench
2. Corten steel
3. Ground
4. Concrete containment
5. Sand
6. Gravel
7. Artificial stone path
8. Drainage system

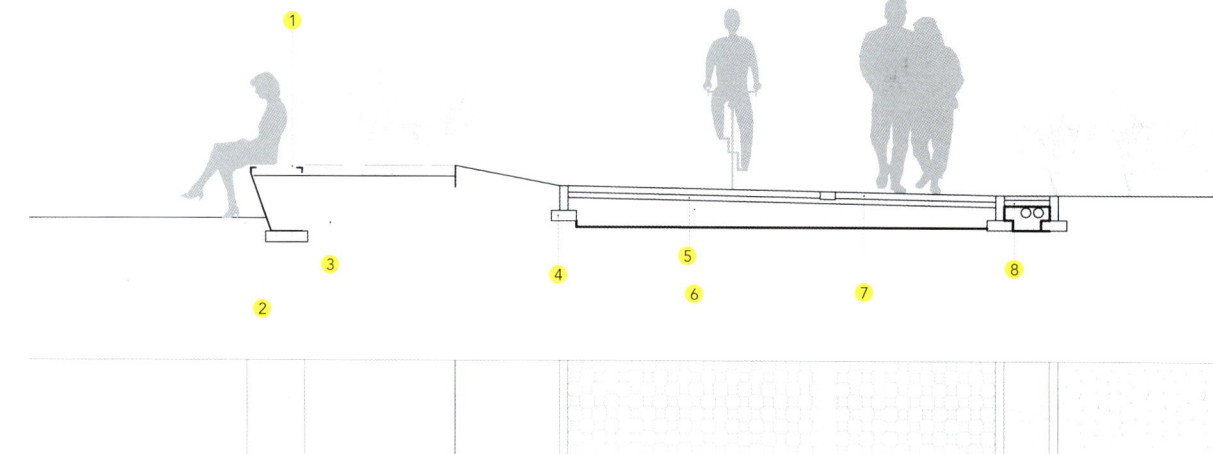

(03) A global suggestive image of the path: the bike path project transforms a disused railway into an urban space that increases the sustainable mobility

(04) The project strategy is to consider the landscape as the most important influence on the characterization of the path

(05) The vegetation underlines the principal public places

(06) A view to the sea, beyond the wooden sofa

(07) The wooden bench near the restaurant point: the vegetation defines the volume between the artificial stone path and the socialization area that is realized with more dynamic materials, such as gravel and sand

(08) The artificial hills are created by recycling the refuse of excavation, in according to the purposes of a sustainable urban design

Northwest Arkansas Razorback Regional Greenway

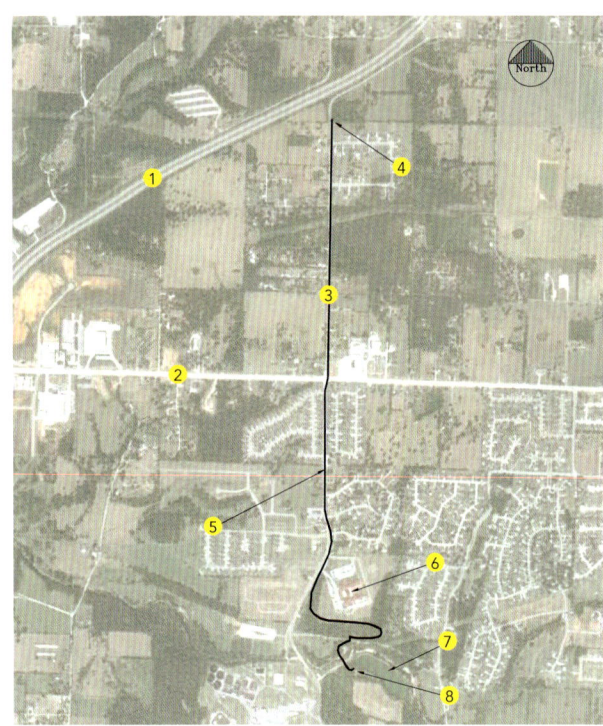

Vicinity map

1. Interstate 540
2. Wagon Wheel Road
3. Silent Grove Road
4. Start trail section
5. Trail section
6. JB Hunt School
7. Lake Springdale
8. End trial section

● Location / Benton and Washington Counties, United States ● Area / 128,036 acres (51,813 hectares) ● Completion / 2015 ● Landscape design / Alta/Greenways, LLC ● Photographer / Charles A. Flink ● Client / Walton Family Foundation and Northwest Arkansas Regional Planning ● Budget / US$25,000,000

The Northwest Arkansas Razorback Regional Greenway is a unique regional urban greenway trail project in the United States. The goal of the project is to link the six communities of Northwest Arkansas (Bentonville, Rogers, Lowell, Springdale, Johnson, and Fayetteville) via a 36-mile-long (58-kilometer-long), primarily off-road, 10- to 12-foot-wide (3- to 4-meter-wide), hard-surfaced trail. It is a project that has been years in the making. Northwest Arkansas has become more urbanized during the past two decades, and the trail route snakes its way through the existing urban fabric of the six communities, taking advantage of undeveloped parcels of land, and, in some cases, restoring and revitalizing underutilized urban landscapes.

The design challenges of the Razorback Greenway were numerous. The project spans six different jurisdictions and required the design team to obtain separate construction approvals from each of the communities. Topography is one of the most important physical challenges of the project. The karst landscape of Northwest Arkansas forced the design team to select a route and alignment for the trail that minimized the impact on native environments. Consequently, there are significant portions of the trail that are built parallel or in close proximity to existing roads. The greenway includes public trailheads built at strategic locations that facilitate ease of access and use, and some segments of the greenway can be lit for night-time use. It is a high-quality, paved, all-weather facility that can be used year-round.

The Razorback Greenway is the human-powered equivalent of Interstate Highway 49 (restricted to automobile use). It follows roughly the same route as I-49, but it is off-road, and supports non-motorized travel linking residents to dozens of key community destinations, including the six downtown areas, three major hospitals, twenty-three schools, the University of Arkansas, shopping areas, residential communities, and popular attractions. In addition to providing access to safe and attractive places to cycle, walk, hike, jog, skate, and enjoy water-based trails, the Razorback Regional Greenway also serves as an easily accessible outdoor laboratory for the thousands of schoolchildren in the area, as well as undergraduate and postgraduate students at colleges and universities in the area.

01 Aerial view of Rogers, Arkansas, showing route of the Razorback Greenway
02 Cyclists approach a bridge spanning Spring Creek in Springdale, Arkansas

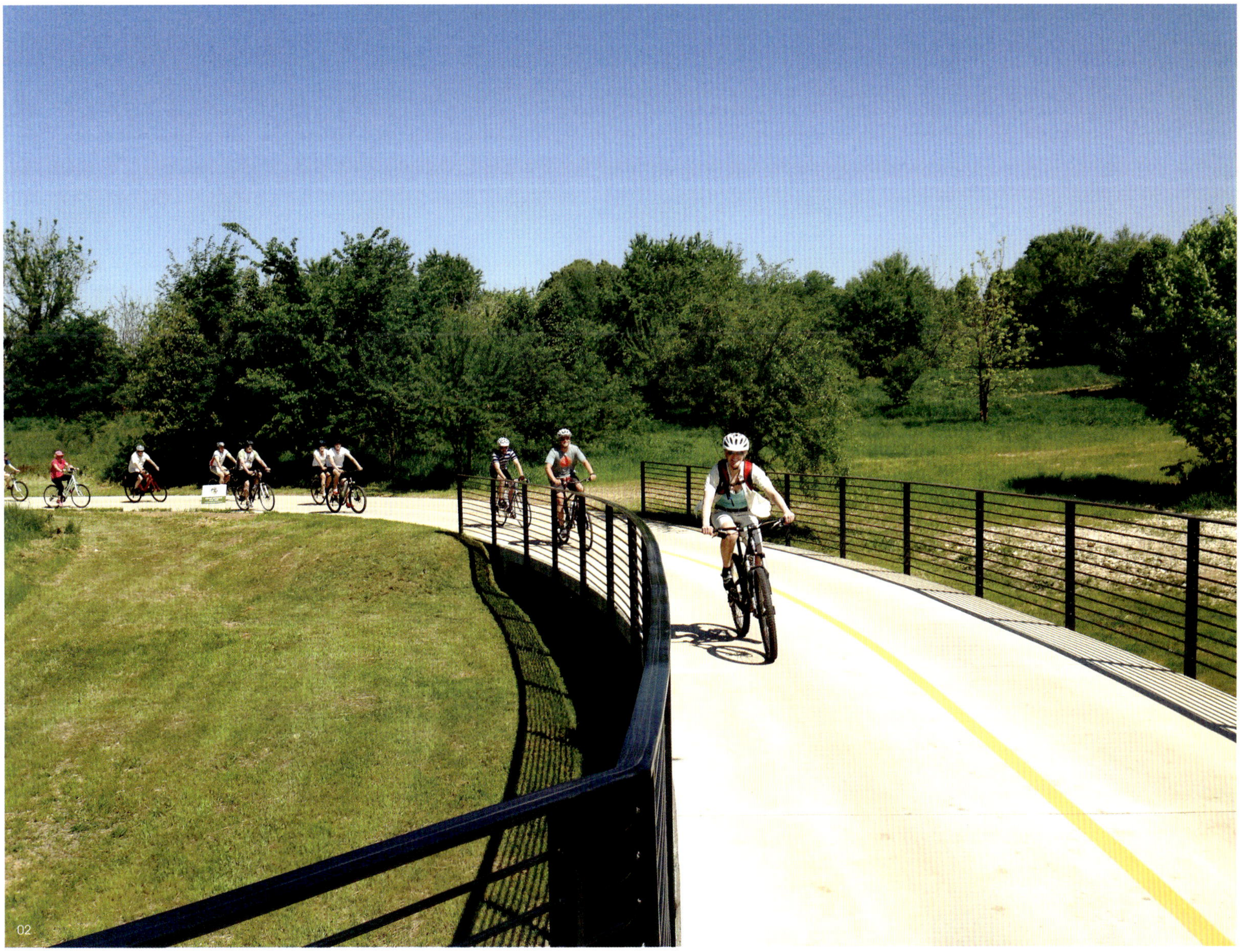

③ Razorback Greenway along Spring Creek, Springdale, Arkansas
④ Razorback Greenway adjacent to Magnolia Gardens, Springdale, Arkansas
⑤–⑥ Looking north on Boardwalk Trail, downtown Springdale, Arkansas
⑦ Boardwalk trail in downtown Springdale, Arkansas

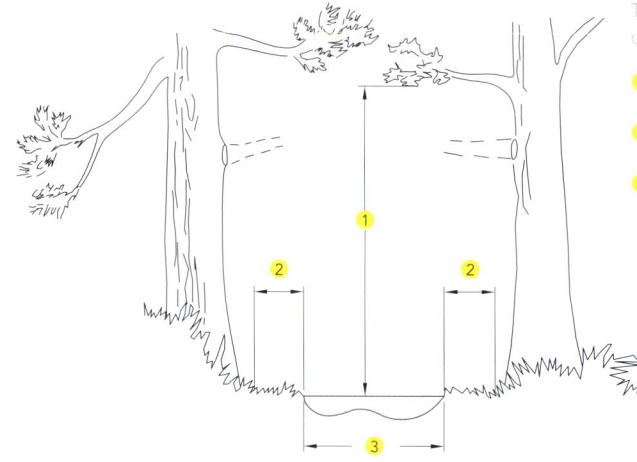

Typical tree and vegetation clearing detail

1. 10 feet (3.04 meters) minimum, vertical clearance
2. 4 feet (1.21 meters) minimum, clearance from edge of trail
3. Trail bed (width varies)

Roadway crosswalk pavement marking detail

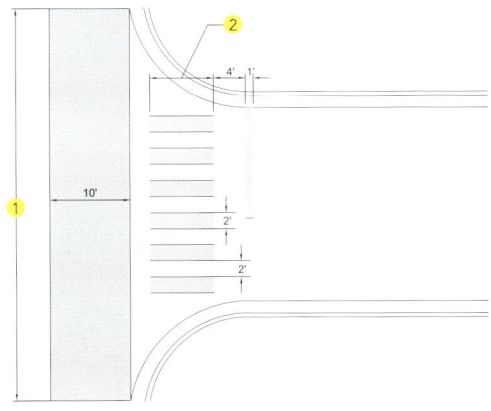

1. Variable length
2. Match trail width or 6 feet (1.83 meters) if next to bicycle track

Notes
a. All crosswalks and stop bars to be white thermoplastic material.
b. Crosswalk to be centered on handicap ramp and be perpendicular to street centerline
c. Where pedestrian and bike crossings exist in the same place, bike track markings shall be a green, solid, epoxy-modified acrylic water-borne coating, and pedestrian markings shall be white thermoplastic material

Cross-section of concrete trail

1. 12 feet (3.67 meters) typical
2. 2 feet (0.61 meter) minimum
3. Two percent cross slope
4. 6:1 slope max
5. Conduit for future electrical connection
6. Concerete pathway
7. Aggregate base course
8. Compacted subgrade
9. Undisturbed subgrade

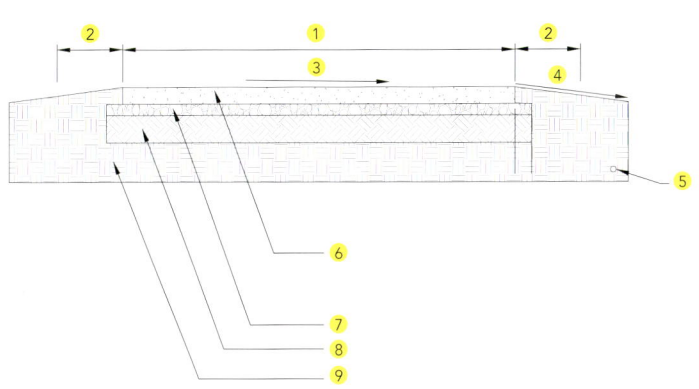

08 Crossing system on Pump Station Road, Springdale, Arkansas
09 180-foot (54.86-meter) span prefabricated steel bridge over Spring Creek, Springdale, Arkansas
10 A family enjoys picnic and views at trailhead in Rogers, Arkansas
11 Cyclists on the Razorback Greenway, Rogers, Arkansas
12 Mile post signage for Razorback Greenway, Springdale, Arkansas

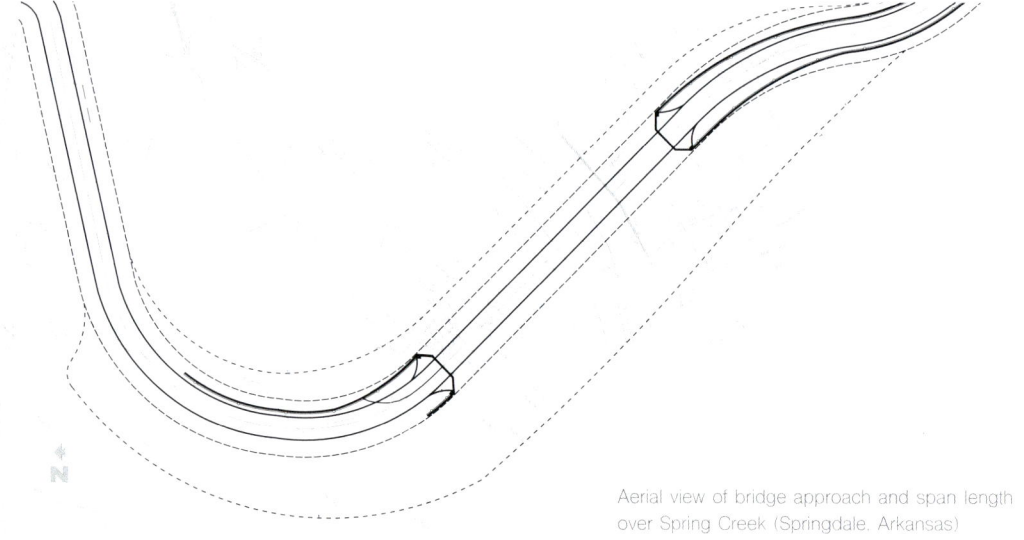

Aerial view of bridge approach and span length over Spring Creek (Springdale, Arkansas)

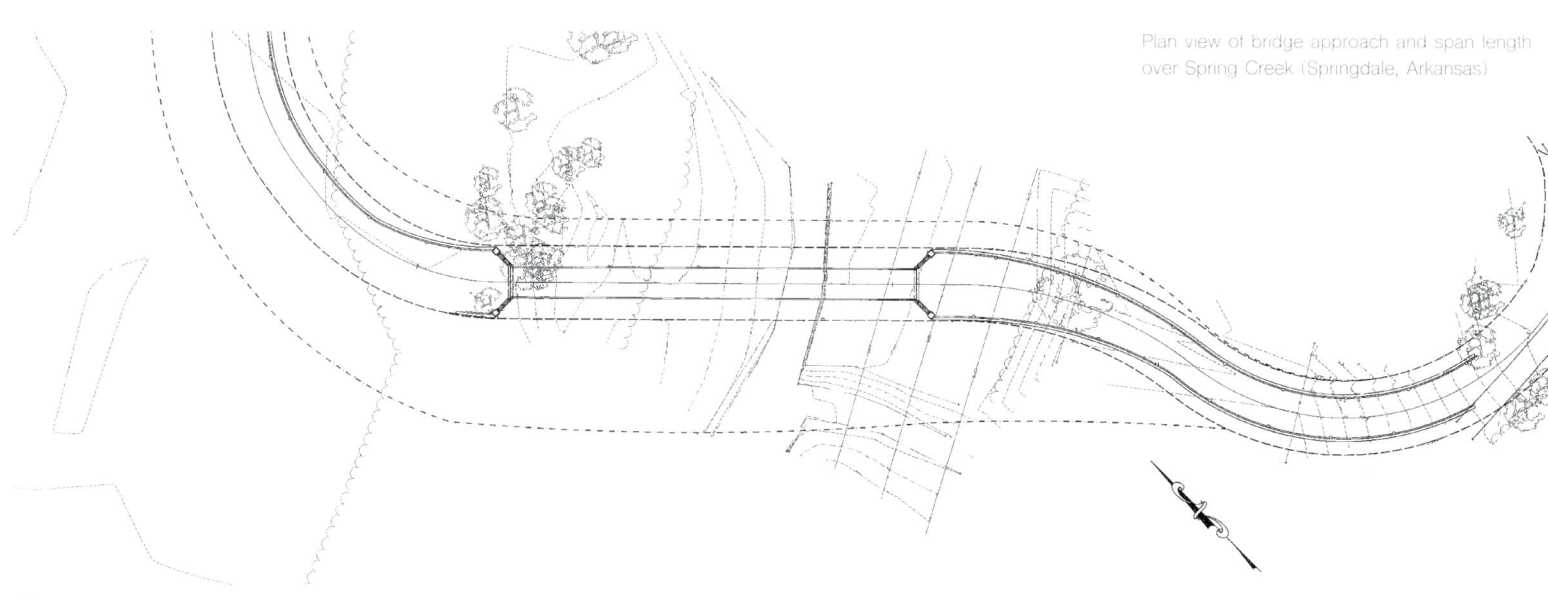

Plan view of bridge approach and span length over Spring Creek (Springdale, Arkansas)

Cross-section view of bridge span over Spring Creek, Springdale, Arkansas

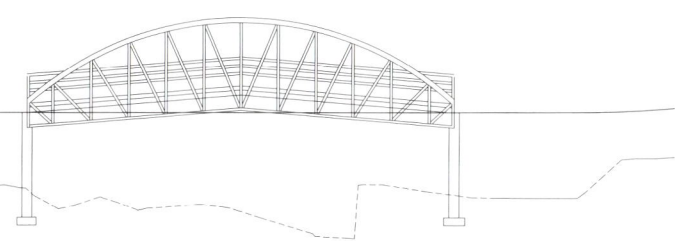

⑬ Approach railings to bridge spanning Spring Creek, Springdale, Arkansas
⑭ Razorback Greenway trail and fence system adjacent to farmland, Springdale, Arkansas
⑮ Razorback Greenway trail in Compton Gardens, Bentonville, Arkansas

Te Ara/Alexandra Stream bike path

Rosedale Park | Alesandre Creek & Omega Reserve | Rook Reserve | Barbados Wetland | Unsworth Drive Forest | Unsworth Reserve

Master plan

Te Ara/Alexandra Stream bike path

● Location / Auckland, New Zealand ● Length / 2.2 miles (3.5 kilometers) ● Completion / 2012 ● Landscape design / BMLA Ltd., Morphum Environmental, JG Civil ● Photographer / Sean Shadbolt ● Client / North Shore City ● Budget / NZ$950,000

The interest in cycling as a sport and pastime for New Zealanders has continued to grow. In Auckland, local government has been active in developing a network of bicycle paths alongside the forgotten and, in many cases, degraded suburban streams. Urban streams in Auckland actually connect many council-owned parks and reserves. The construction of bike paths links up a whole series of what had been disparate and underutilized public spaces. It can also be coordinated with the restoration of degraded streams, wetland enhancement, and the upgrading of existing stormwater ponds adjacent to streams.

The Alexandra Stream bike path on Auckland's North Shore is a good case of this bike path initiative. The Alexandra Stream is a medium-sized urban stream that runs roughly south to north from Unsworth Heights to the Albany Creek, about 2.2 miles (3.5 kilometers). It runs through a number of existing parks and reserves, notably Unsworth Reserve and Rosedale Park.

The designer carried out a number of detailed studies focusing on the specific environmental conditions that the track would pass through, which revealed a complex landscape with overland flow paths, ephemeral streams, fragments of native bush, and a steep topography. This study led to the development of a range of specific interventions that responded to site-specific situations. In some cases, construction techniques that have not been traditionally used in park construction were adapted, for instance, steel sheet piling, a material and technique usually applied to the construction of large-scale infrastructure projects, was used to retain sections of the path. Native bush fragments that would have been destroyed by using conventional retaining walls could be preserved by adopting thin sections of steel sheet and a specific installation technique. Permeable concrete was used for some sections of the path to reduce stormwater runoff and improve water quality. Recycled, crushed, construction concrete was used in drainage trenches. Green Terramesh walls were used to retain the lower sections of paths to ensure the integration of native-plant restoration with the retaining-wall structures. Construction materials were selected for minimal maintenance: tanalized timber bridges, galvanized-steel light poles, and poured concrete seats with a simple board-faced finish were some of the material choices.

The Alexandra Stream bike path has created a new facility and opens up alternative routes for the citizens in the car-dominated suburbs of the North Shore. Apart from the road networks, new path connections can now be made through the suburbs. From sports fields, cafés, gyms, and offices to suburban houses, a new ecology-friendly, safe network is now available for the community. The Te Ara/Alexandra Stream bike path has become a new linear park for Auckland.

Bike path from Paul Mattews Drive to SH 18

1. Poured concrete seat
2. Information board
3. Steel-sheet-piling retaining wall
4. Flax under bridge
5. Timber bridge
6. Timber deck
7. Overflow path under deck
8. Mass hebe planting
9. Mass toitoi planting in front of engineered earth retaining wall
10. Wetland planting around culvert outlet

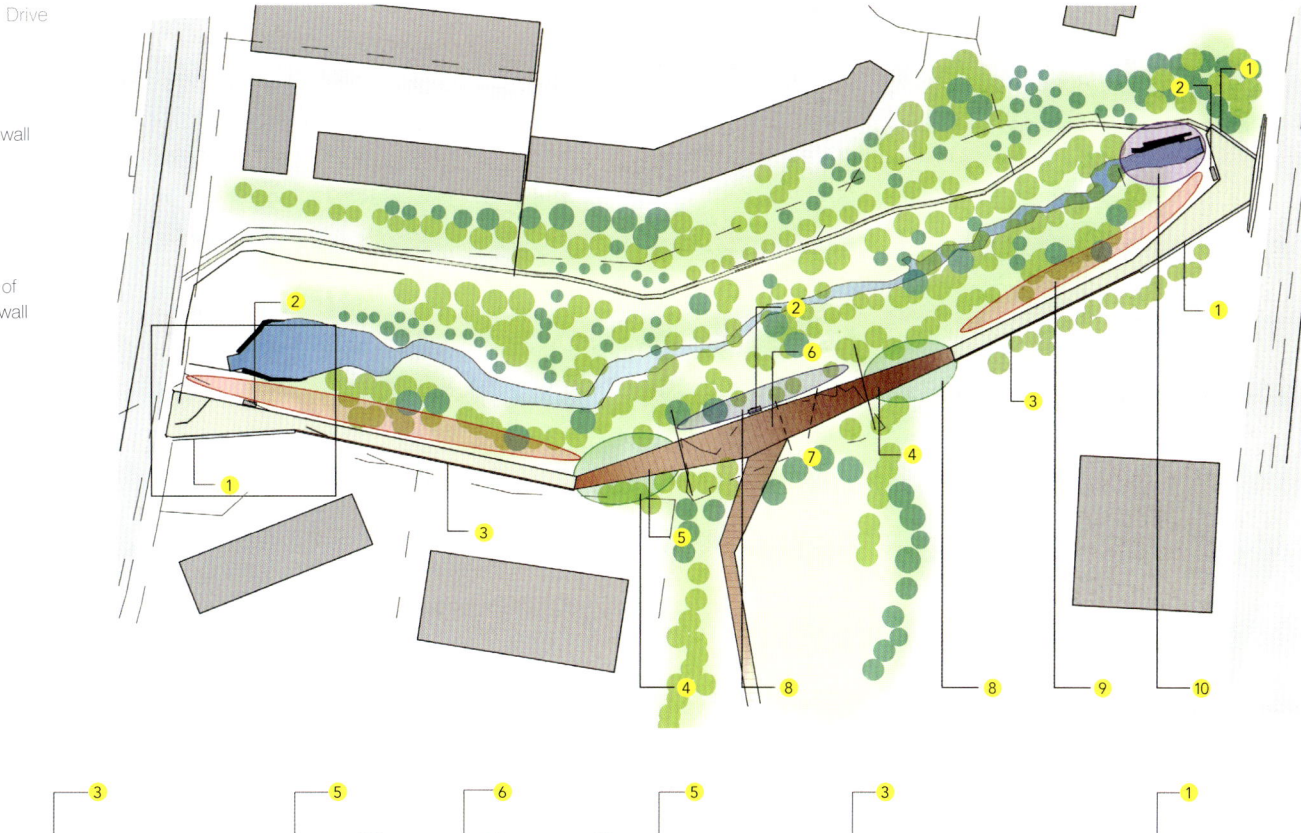

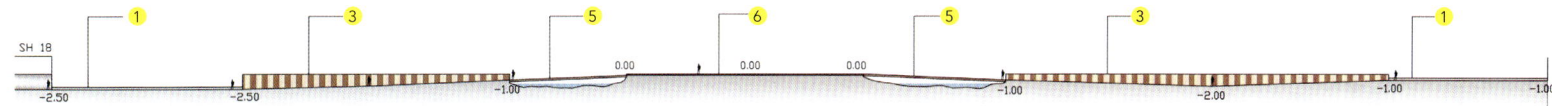

01 Bike path through Unsworth Reserve

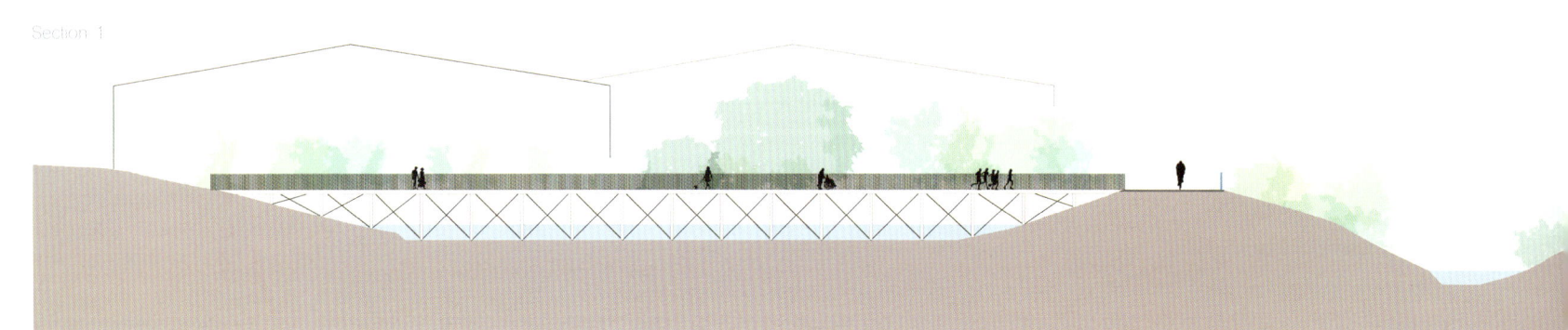

Section 1

02 Bridge over Omega stormwater pond
03 Bike path over Omega stormwater pond from Omega Street
04 Bridge over Omega stormwater pond from bike path

05–06 Steel-sheet-piling retaining wall
07 Timber seat and bridge
08 Tunnel with new safety lighting
09 Bike path boardwalk through existing pine forest

Section 2

The Luchtsingel

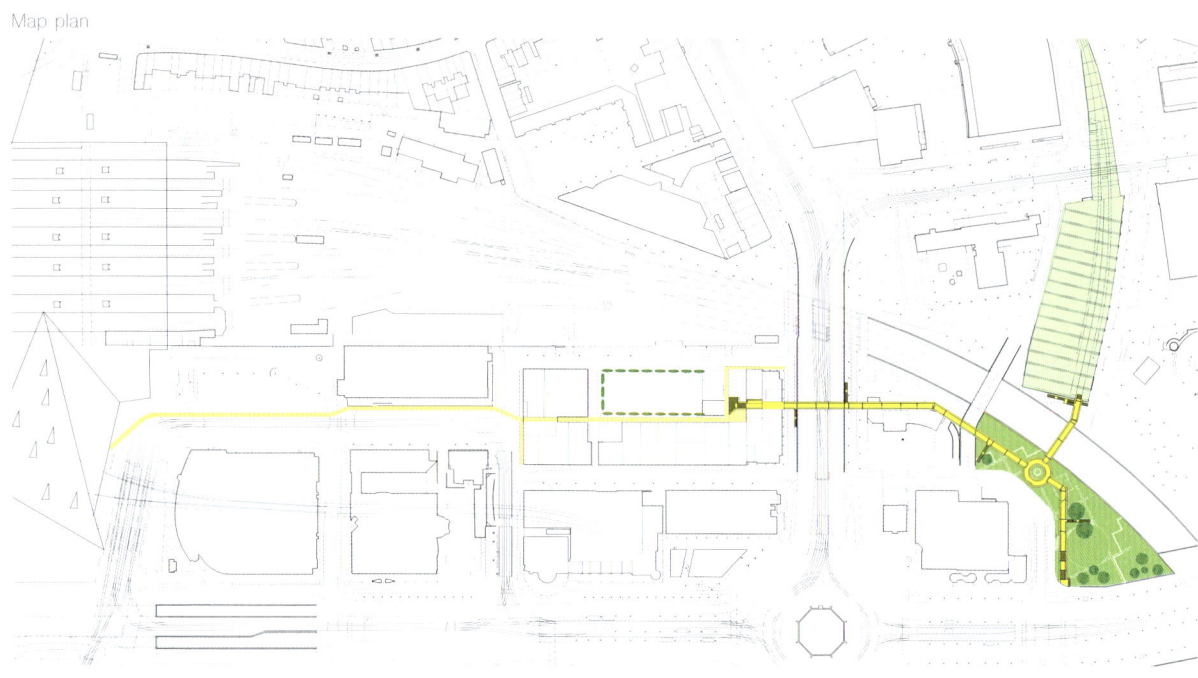

Map plan

● Location / Rotterdam, Netherlands ● Length / 0.25 mile (400 meters) ● Completion / 2015 ● Landscape design / ZUS ● Photographer / Ossip van Duivenbode ● Client / Rotterdam Township ● Budget / €1,500,000

The Luchtsingel is a 1312.34-foot-long (400-meter-long) wooden bridge for pedestrians, which connects the center and north of Rotterdam. It is the world's first piece of public infrastructure to be accomplished mostly through crowdfunding. Together with the new public spaces, including Schieblock (an office building for young entrepreneurs), Delftsehof (a vibrant nightlife area), Dakakker (a roof garden), Pompenburg Park, and the Hofplein Station Roof Park, Luchtsingel has contributed to a "three-dimensional cityscape."

In 2011, it was announced that a planned office development in Rotterdam Central District had been canceled, leaving many existing office spaces vacant. In order to revitalize this area, the designer conducted several redevelopment projects including Schieblock, Delftsehof, Dakakker, Pompenburg Park, and the Hofplein Station Roof Park. The Luchtsingel is one of the significant projects, which runs throughout these other projects as a unifying factor linking these varied and new public spaces and returning this former heart of Rotterdam back to being green and livable. To create a unifying image with other projects, the Luchtsingel is a bright yellow color, which is consistent with the yellow lines painted on the outside ground of Schieblock and a walking path in Delftsehof. This striking color makes the bridge stand out among the surrounding gray buildings. The bridge is extended in three directions with a roundabout in the center, which also serves as a dining place where people can enjoy the view of the city. Another conspicuous feature of the bridge is its plank handrails. The handrails are constructed in an inward manner with the names of the crowdfunders carved on it. Each funder can find a plank with their own name. Because of its length and complex structure, the bridge was constructed and assembled in segments.

By increasing accessibility for pedestrians, the bridge ensures synergy between the various sites. It is now usual to walk from the Station Quarter to the north and to the Laurenskwartier via Pompenburg. This distinctive pedestrian-friendly connection not only provides an alternative for public movement, but also gives the area a unique position in Rotterdam's urban fabric.

01 Bird's-eye view of the bridge from the east

Master plan

02 Bird's-eye view of the bridge from the west
03 The bridge is extended out in three directions with a roundabout in the center
04 The bridge connects Schieblock and Dakakker
05 The inner side of the bridge is like a train carriage
06 The yellow train on the railway as a background
07 Every plank is carved with the name of a crowdfunder

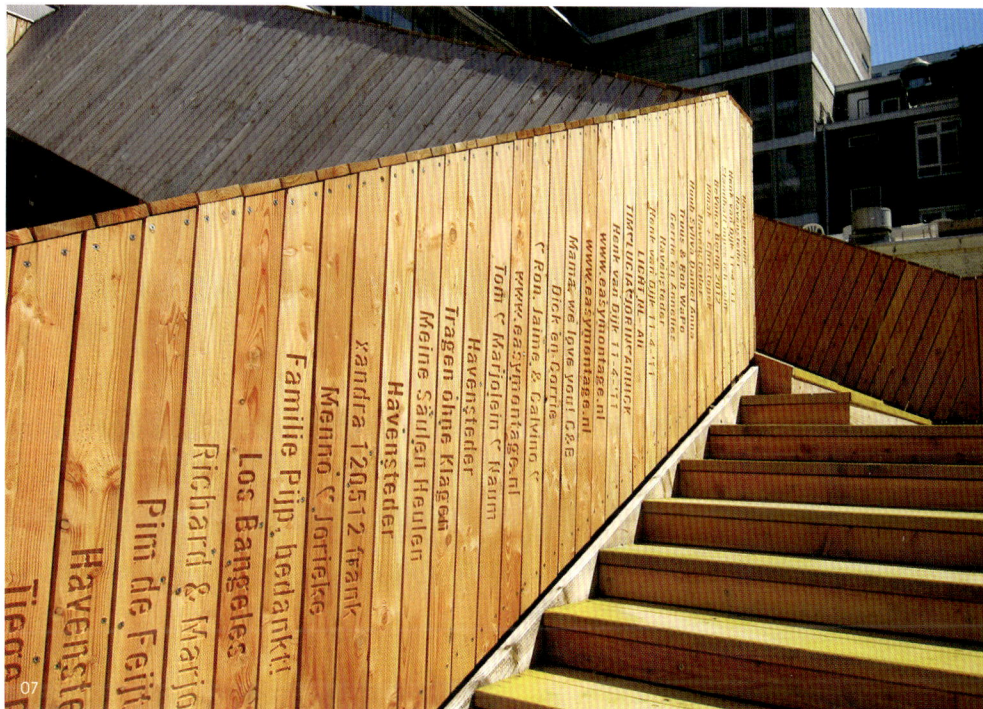
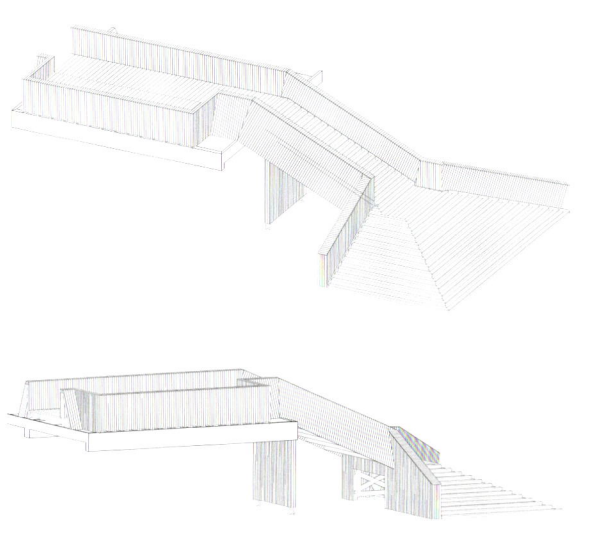

Delftsehof staircase

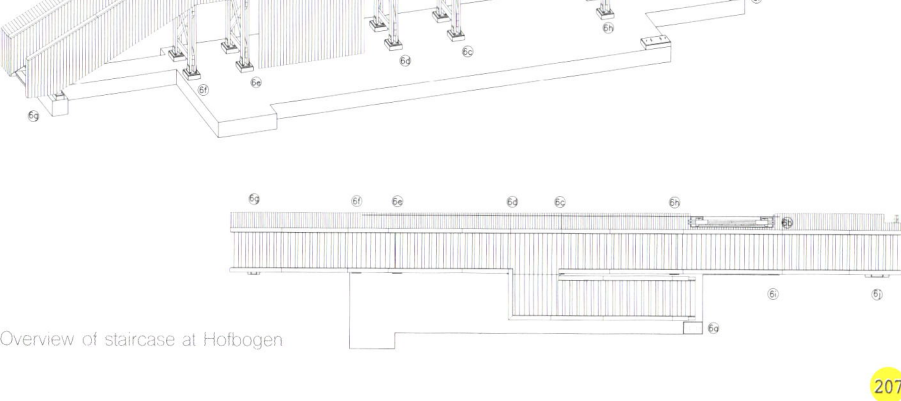

Overview of staircase at Hofbogen

08 The roundabout also serves as a dining place
09 A crowd of visitors standing at the roundabout
10 – 12 The bright yellow color of the bridge forms an integrated landscape with the light in the Schieblock and the yellow line painted on the pedestrian path in Delftsehof

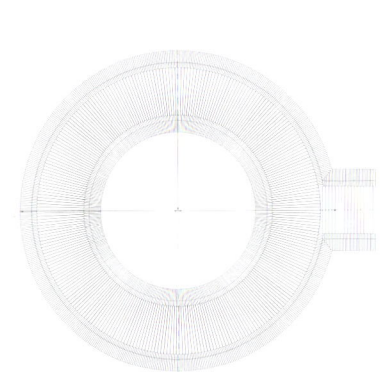
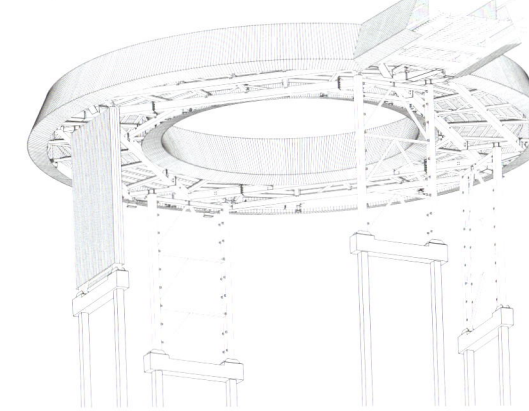
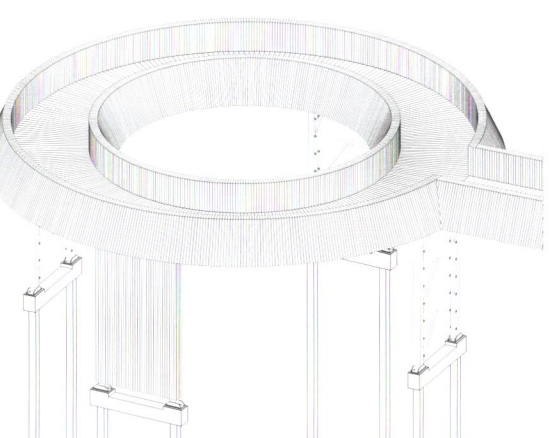

Overview of the roundabout

Bicycle Underpass, Haarlem

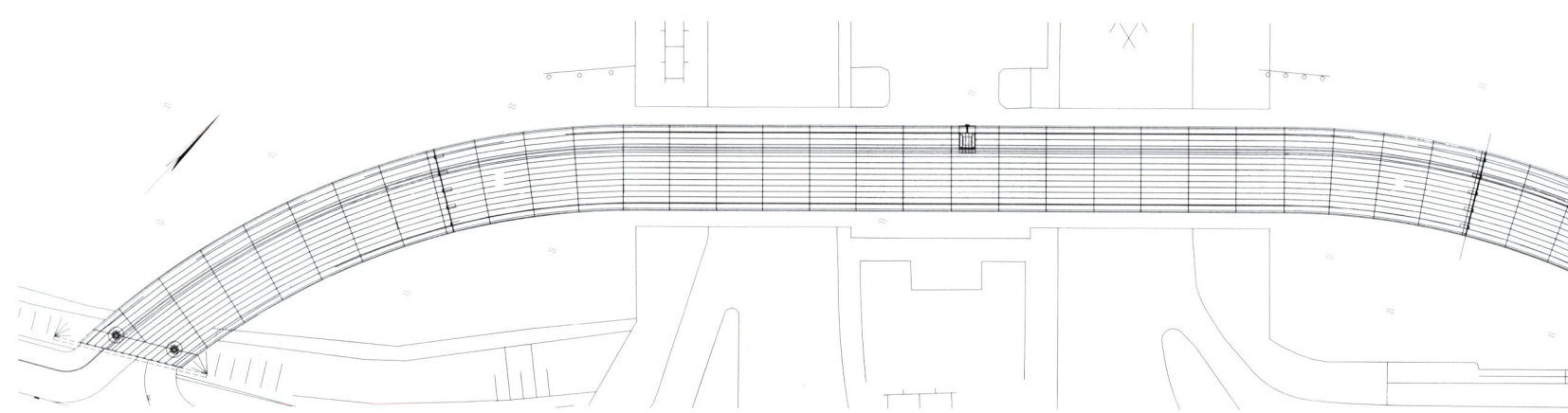

Master plan

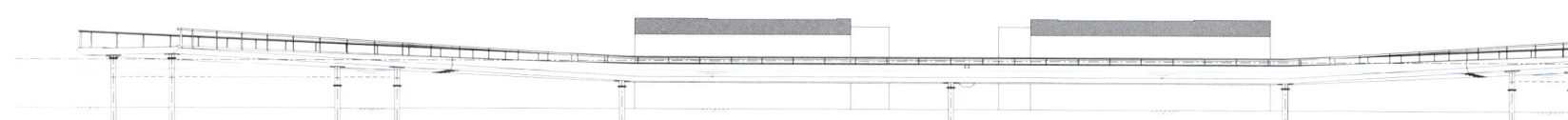

Elevation

● Location / Haarlem, the Netherlands ● Length / 369.6 feet (110 meters) ● Completion / 2012 ● Architect / ipv Delft ● Photographer / HenkSnaterse, Gerhard Nijenhuis ● Client / Haarlem City Council ● Budget / €977,000

The City of Haarlem wanted to make it easier for cyclists and pedestrians to cross the motor traffic road N205 and continue their journey on the main bike route along a river bank. But there was no space for an overpass, so the designer came up with the plan to lead the bike route through the river under the approach span of the two existing drawbridges for the arterial road.

Since there is not enough clearance under the drawbridges, the new bicycle underpass is a 16.40-foot-wide (5-meter-wide) bridge that partially lies about 9.84 inches (25 centimeters) below water level. The railing clearly shows where the bridge deck goes below water level, as it also acts as a barrier. The slope and length of the underpass have been chosen to ensure a good balance between user comfort and building costs. The entire steel structure was prefabricated elsewhere and then transported to the site over water. Despite its weight of over 100 tons, the structure stayed afloat due to its box-shaped, airtight construction. Once in position, concrete counterweight was added, which made the structure sink to its required level on top of the previously placed pole foundation.

The underpass connects well to the existing bike paths, and as it doesn't physically touch the existing traffic bridges, it doesn't interfere with their function, and allows the water to flow freely, which prevents dirt piling up. An integrated water pump enables rainwater to be pumped back into the surrounding waters.

Cyclists and pedestrians who want to cross the intersection now can simply take the underpass that leads them quickly and easily to the other side. The underpass is well lit at night, highlighting the presence of the Spaarne waterway.

01 The rust bridges above the underpass are on a busy entrance route
02 The underpass cuts through the water

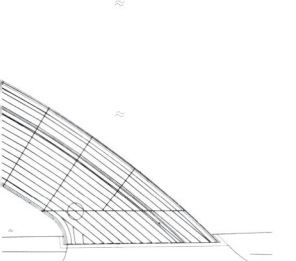
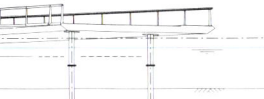

03 Bicycle Underpass, Haarlem rendering
04 A gentle slope leads cyclists down the underpass
05 The bridge deck is illuminated by the handrail lighting at night
06 The width and height of the underpass ensure a feeling of safety

1. Minimal clearance (8.20 feet [2.5 meters])
2. Gradient 5 percent
3. Safety margin area
4. Center line
5. Connection always below water level
6. Rainwater drainage

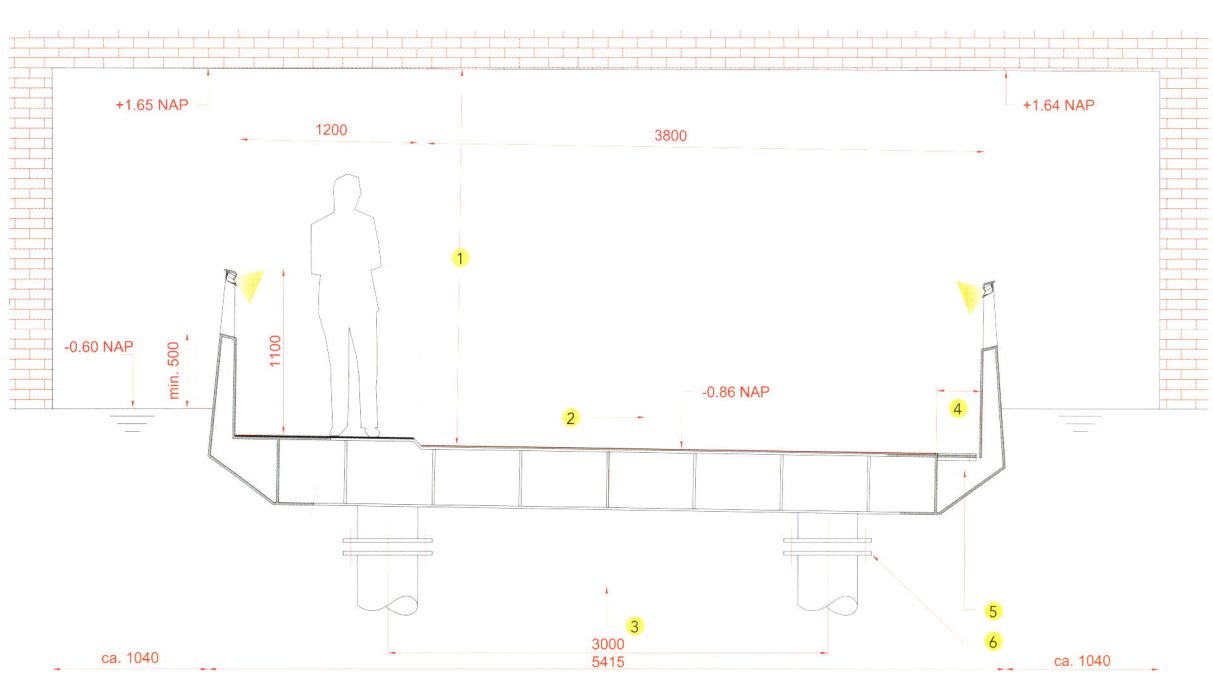

Section

El Valle Trenzado

Master plan

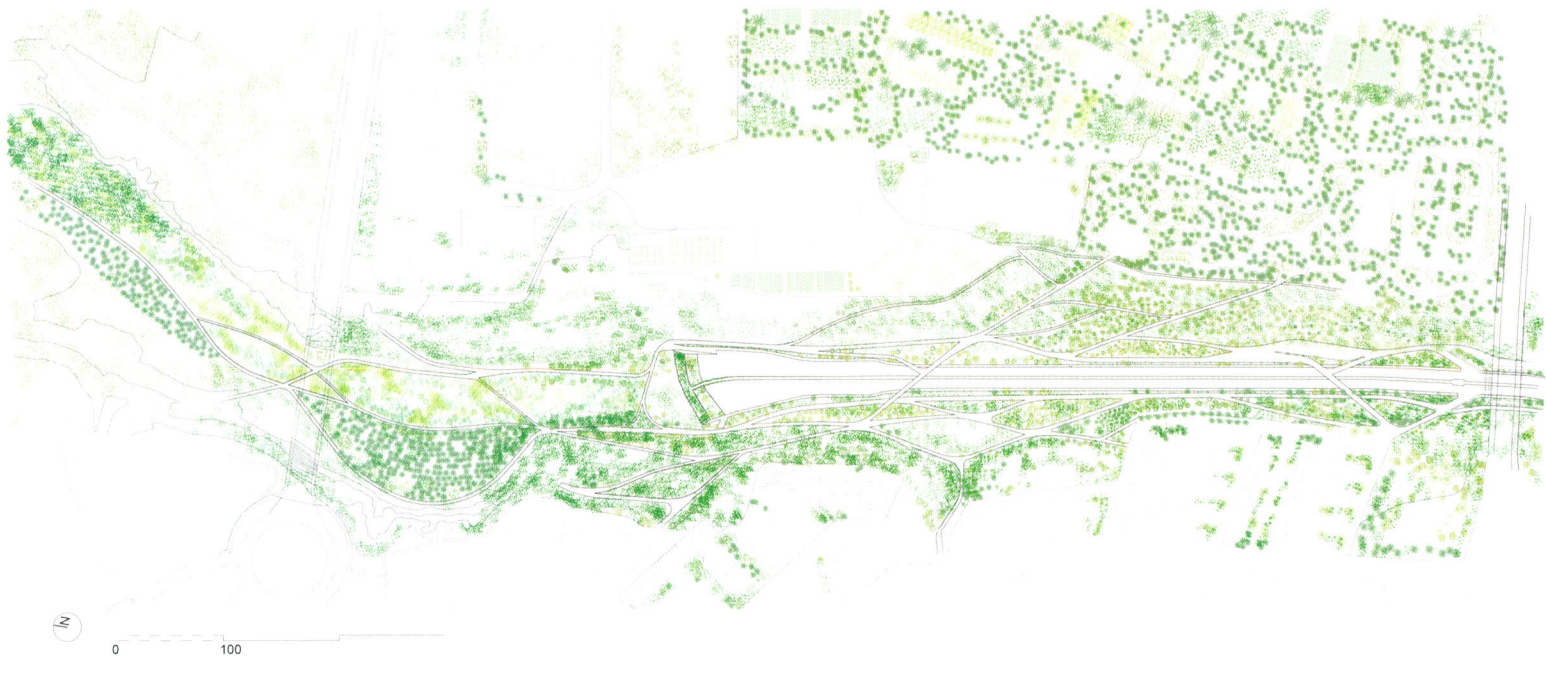

- Location / **Elche, Spain** ● Area / **28.41 acres (11.5 hectares)**
- Completion / **2013** ● Landscape design / **Grupo Aranea Ltd.**
- Photographer / **Elche City Council** ● Client / **Elche City Council** ● Budget / **€2,382,000**

El Valle Trenzado aims to recover the pedestrian's traffic footprint. The channeling works that were made in the 1970s definitely skewed Elche's Canyon. For years, the citizens of the city went through this place, hardly enjoying it.

This landform, which reaches 131.23 feet (40 meters) deep, gives the city a chance to travel in a few minutes to an area of high environmental quality. The designer proposed a system that can be adapted to the geographical and administrative complexities of the place, negotiating with all agents who claimed the space as their own to make it more accessible and increase the public use of the steep slopes. The system began with a month-long consulting process, allowing citizens and groups to participate. The project then incorporated the main points, areas of special interest, the most requested routes, and the large range of desired uses into the project design. After this first step, the system was tested again, managing the complex and contradictory relations between multiple administrative stakeholders responsible for the place.

The system aims to give voice to the Vinalopó River. The sinuous roads have no relation to the orthogonality of the city. These squiggly lines shape the slopes, offer more comfortable routes, and incorporate the possibility of crossing the river. The roads that float on the untouchable riverbed of concrete blur the distinction between bridge and path, becoming a thought solved by a material abstraction. A single bridge becomes a network of trails, which fold, bend, stretch, tighten, disperse, curve, and, of course, twist. The vegetation is also a construction material. Native species and trees help to build the shadow of the forest. The system is based on experience, understanding, and empathy, connecting the valley and reducing infrastuctures to a minimum. Over time, they will allow the flourishing of urban activities.

01 Cyclist crossing El Valle Trenzado
02 Rendering of the typology of the area
03 Model of Elche's Canyon together with the City Hall

04 A rendering to show the bridge connecting with the local paths in both shape and color, which has blurred the difference between bridge and path
05 The bridge lands on a path with folded turn just like the shape of the bridge
06 The bridge floats on the untouchable riverbed of concrete
07 Native species and trees help to build the shadow of the forest

06

07

08 The bridge with the residential buildings in the background
09 Detail of the bridge
10–12 Different views of different uses and connections of the bridge

Glacier Skywalk

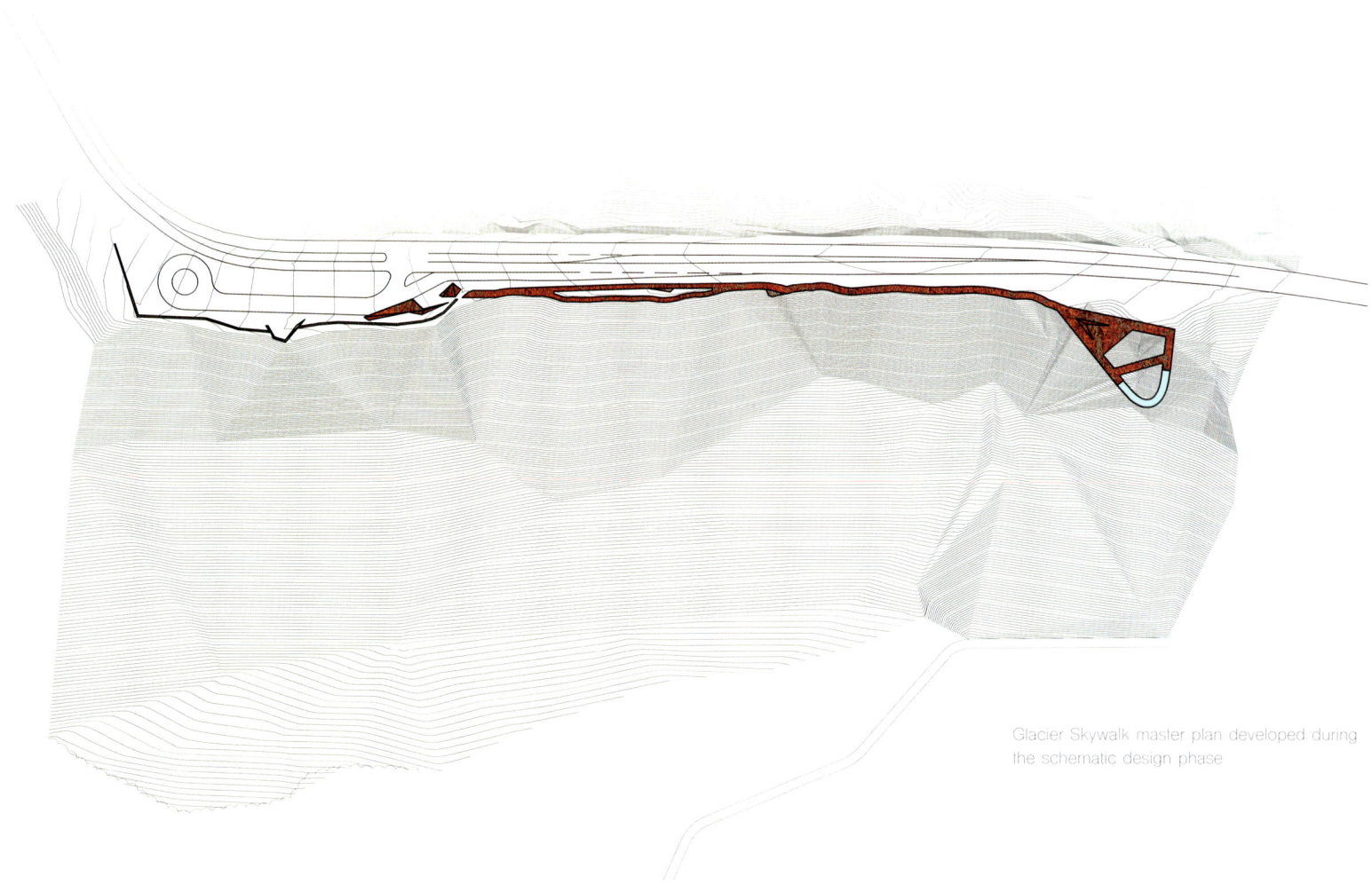

Glacier Skywalk master plan developed during the schematic design phase

● Location / Alberta, Canada ● Length / 1476.4 feet (450 meters) ● Completion / 2014 ● Architect / Sturgess Architecture ● Structural engineer / Read Jones Christoffersen ● Photographer / Robert Lemermeyer ● Client / Brewster Travel Canada ● Budget / CA$21,000,000

The Glacier Skywalk is an interpretive walk, carved and folded into the mountainous landscape of Jasper National Park in the Canadian Rockies. The Corten steel and glass structure cantilevers extend outward, overlooking the Sunwapta Valley and facing the Athabasca Glacier.

The walkway is based on the concept of cropping out from the landscape, creating an experience of a natural extension of the land. The designer wanted to give people the opportunity to get out of their car to experience this incredible landscape in a way that would provide a cerebral connection to the changing natural environment. The design is founded on the idea of a mountainside outcropping that exists as an organic extension of the landscape.

The walkway begins at a kiosk next to a drop-off area and continues along a trail with different zones dedicated to the ecology, geology, and history of the landscape. The parabolic outlook is located at the far end of the walkway, where it cantilevers 98.43 feet (30 meters) from the rock face with a glass floor of tempered and heat-strengthened glass, revealing an unobstructed view below. The cantilever is the result of an engineering technique taking advantage of a balance formed by opposing tension and compression components, and thereby eliminating the need for a more traditional superstructure of pylons and cables above the outlook. Corten steel was selected for its weathering properties and the relationship to the changing mountain face that it represents. The materiality of the outcropped steel will age, and Athabasca Glacier will shrink. Both will mark time with change.

01 Artist rendering of the complete interpretive pathway set along the mountainside
02 View from the ticket kiosk looking south along the glacier-formed valley

03 The geology interpretive exhibit emerges from the mountainside as visitors make their way to the outlook

04 The architectural language of the ticket kiosk strikingly emphasizes the fractal nature of the surrounding landscape
05 View from the top of the amphitheater looking back along the interpretive pathway

06 Artist rendering of the glass-floored outlook developed during the schematic design phase
07 Upon approaching the Corten-clad outlook, visitors are met with fractal volumes inspired by the surrounding landscape
08 From below, the parabolic outlook appears to defy gravity as it extends 98 feet (30 meters) from the rock face
09 Closer inspection beneath the outlook reveals the compression components used for the structural system
10 View to the west overlooking the neighboring valley and mountainscape as seen from the outlook

Energieberg Georgswerder Horizontweg

Site map

- Location / Hamburg, Germany ● Length / 3018.38 feet (920 meters) ● Completion / 2013 ● Landscape design / häfner jiménez betcke jarosch landschaftsarchitektur gmbh ● Engineering design / Sauerzapfe Architekten, ifb Frohloff Staffa Kühl Ecker ● Photographer / Hanns Joosten ● Client / City of Hamburg—BSU Authority for Urban Development and Environment ● Budget / €4,300,000

With a height of about 134.51 feet (41 meters), the mountain of Georgswerder on the Wilhelmsburg Island of Hamburg is not really a mountain, but in the flat environment of Hamburg, its height is quite remarkable. It seems like a nice green hill with meadows and groves, but there is a big fence around it. It is a landfill whose surface is spotted by hundreds of gun barrels, pipes, gate valves, hydrants, electrical cabinets, and so on. However, it has been transformed into a new structure of wind power, solar power, and so on, from which it gets the name "Energieberg" (Energy hill).

This project aims to add a horizontal walk to offer visitors a place that explains technical features of the landfill and the nature of current as well as new functions of Energieberg. The difficulty is how to transform an inaccessible place into an attraction. The hill's surface is a complete system and a product of many years of research, and also a valuable biotope. The topography and vegetation have already been finished. The project is not supposed to touch the hill, or change or hide the appearance of the surface. The three wind turbines, which were founded with flat circular foundations above the geomembrane, lead the design for the construction. Horizontweg (the horizontal walk) is connected with the ground on the retaining wall in some parts, and floats up to 22.97 feet (7 meters) over the ground in other parts. Curving like a contour line, the trail of Horizontweg crowns the hilltop and offers views in every direction.

This project has transformed the closed landfill hill into a pleasant destination. Horizontweg presents a special experience of a pedestrian trail. It is also used for jogging events, fashion shoots, brand presentations, and so on. It has a bridge that has a circumferential band of PV modules and LED strips on the outside. Every unit of the bridge is movable and re-adjustable, to allow for shifts in the hill. It is a special destination that creates a unique atmosphere at night.

01–02 Bird's-eye views of the walk
03 Southwest curve of the walk

01

02

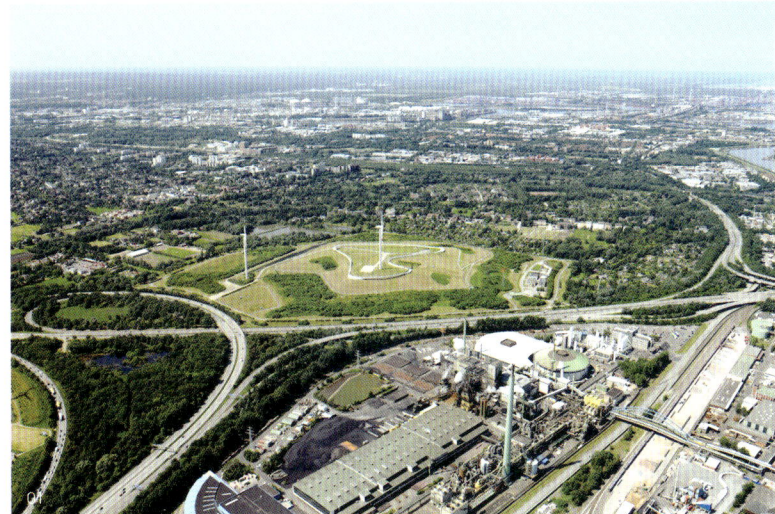

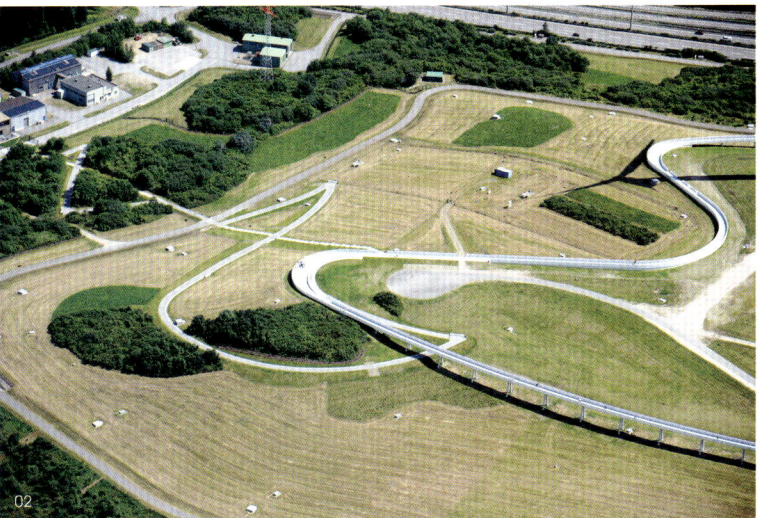

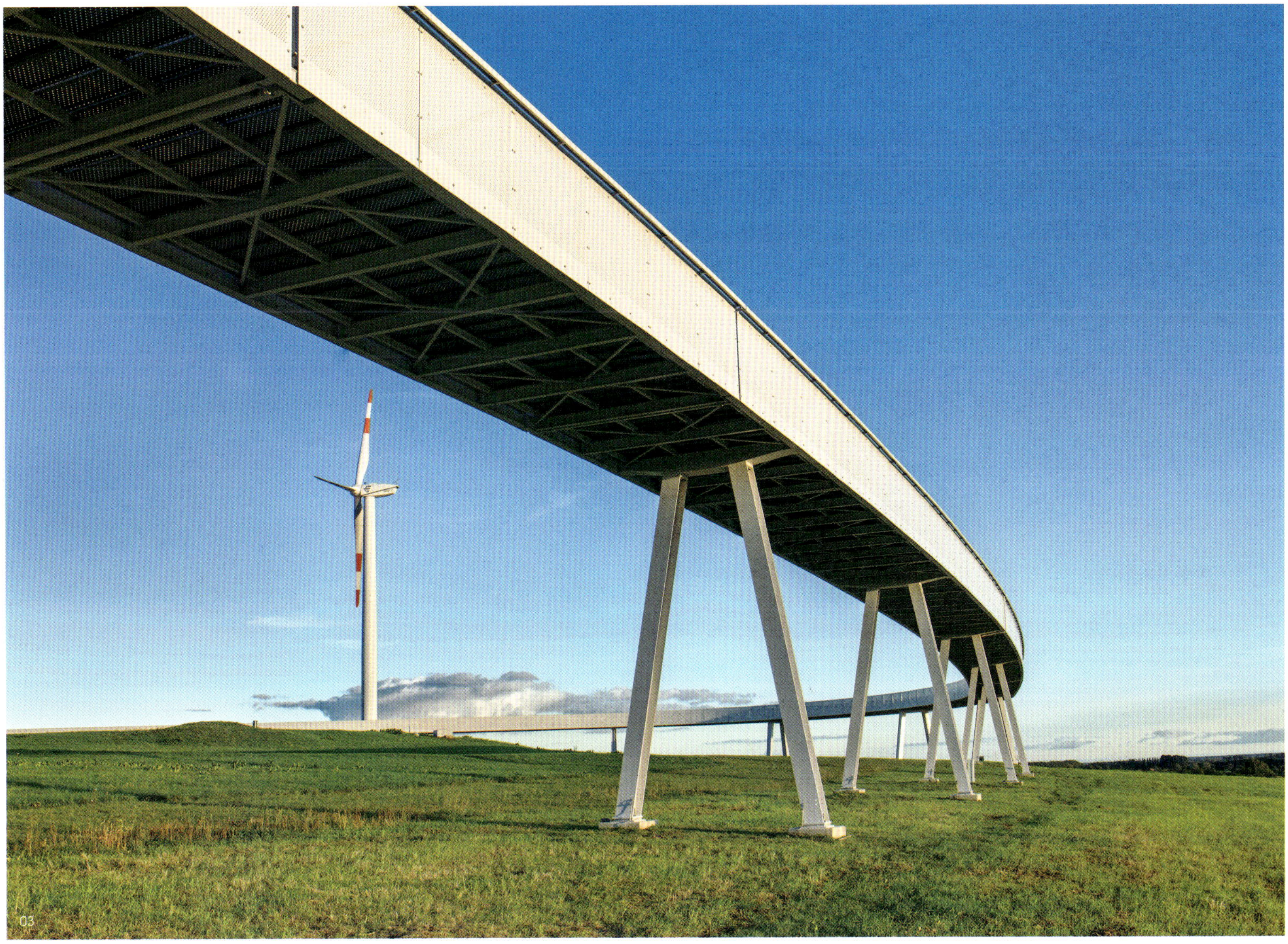
03

04 The walk extends and connects with the skyline in the far end
05 The walk becomes a unique landscape on the hill
06 Inside and outside of the walk with turbines in the background

1. Upper cover substrate
2. Soil filling
3. Railing
4. LED lighting system
5. Concrete angle bracket
6. Soil filling
7. Original ground level
8. Profile sheet grids
9. Lower cover substrate
10. Cover substrate
11. Geotextile
12. Drainage layer
13. Plastic sealing sheeting

Intersection of the horizon walk touching the ground level

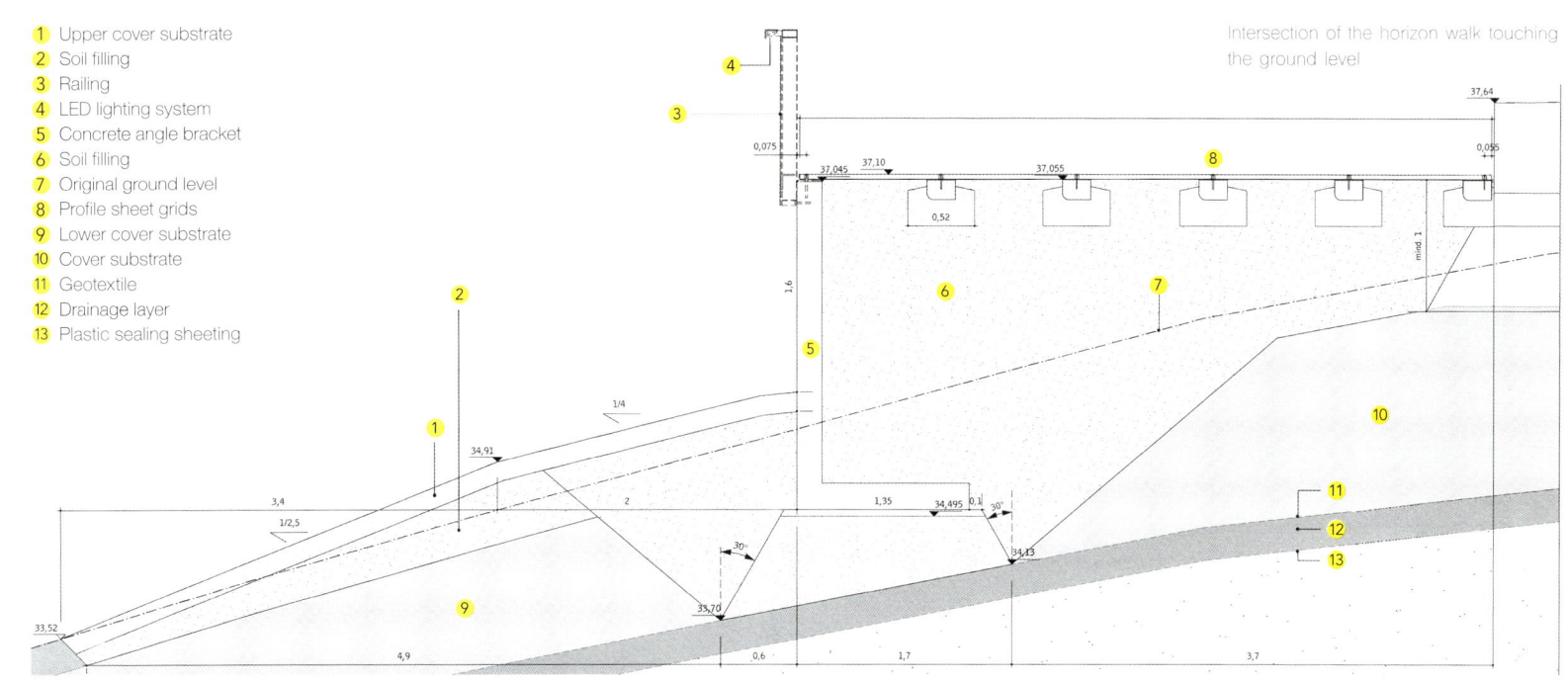

07 The underneath of the bridge
08 - 09 The distant view shows the purpose of the walk: reaching and leaving the hill's surface

Idea of the walk as horizontal landmark

Plan of the northwest curve

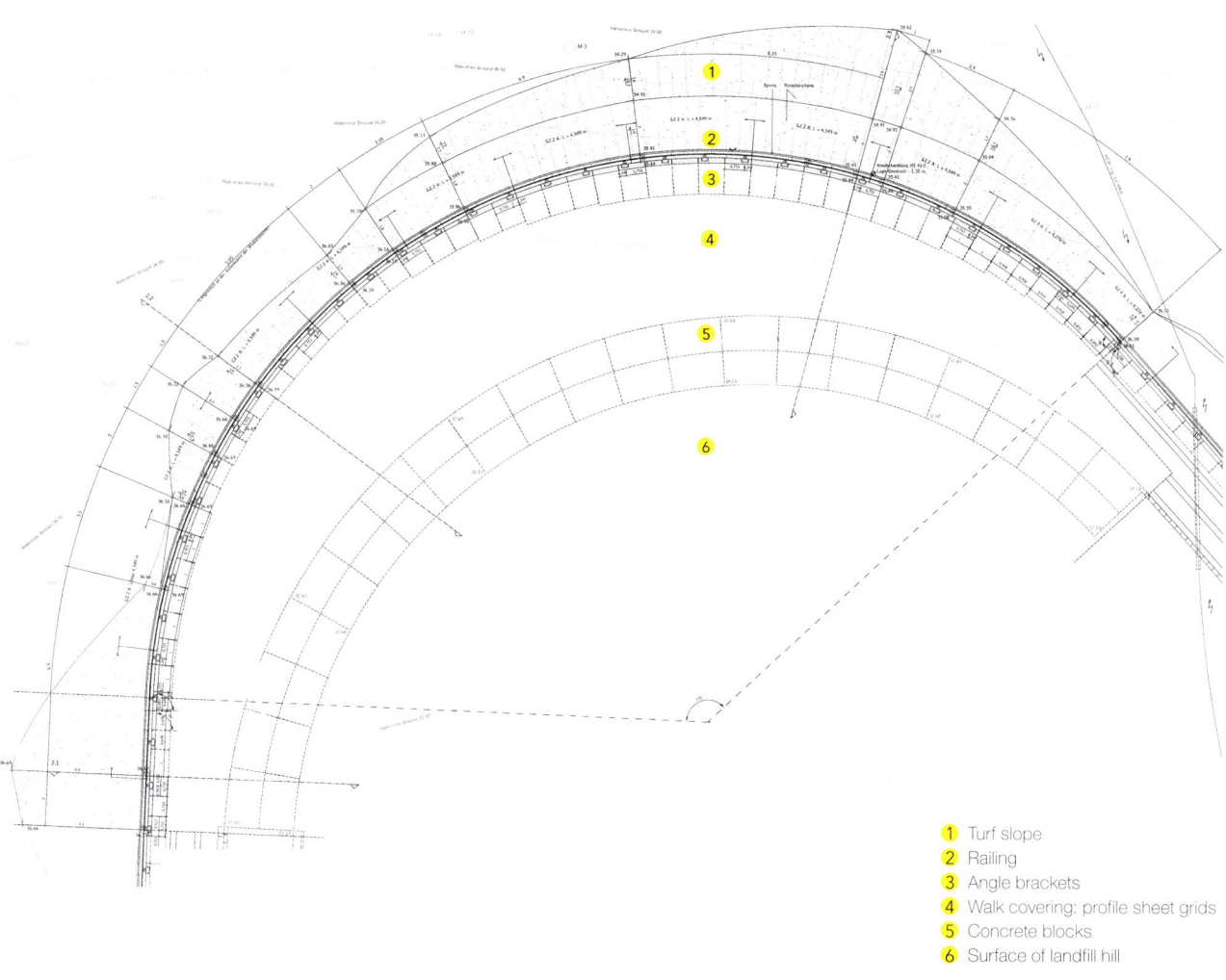

1. Turf slope
2. Railing
3. Angle brackets
4. Walk covering: profile sheet grids
5. Concrete blocks
6. Surface of landfill hill

Intersection of steel column axis 2

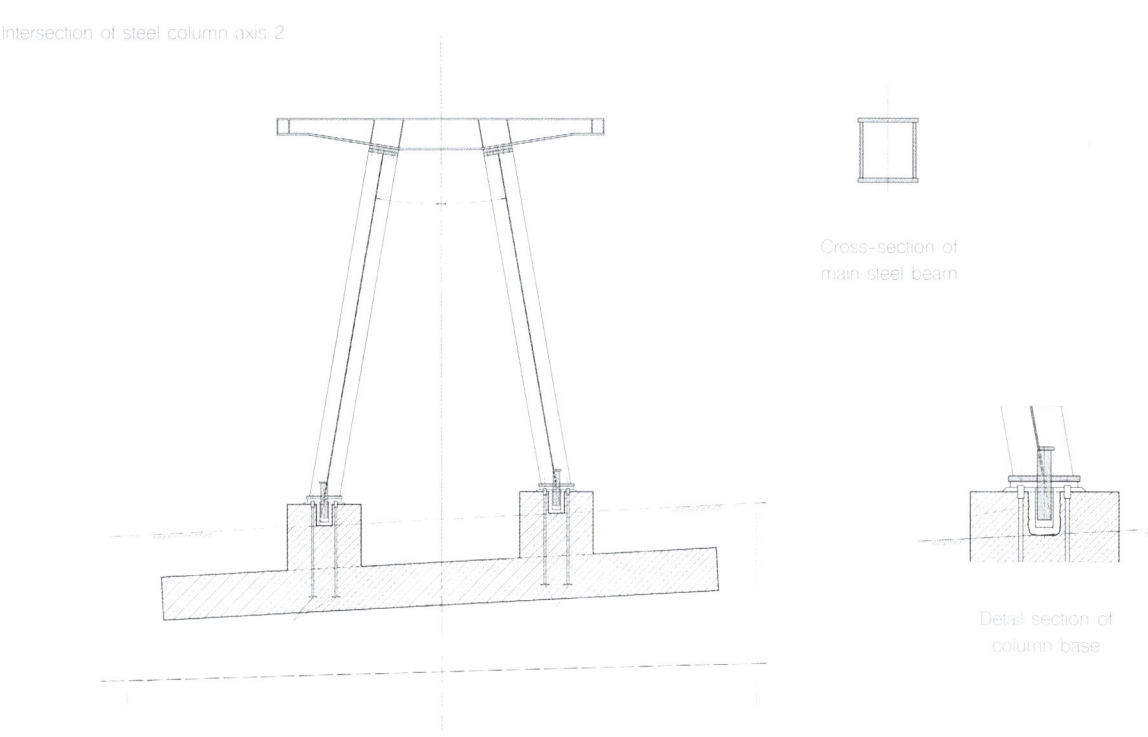

Cross-section of main steel beam

Detail section of column base

Plan of the basis of steel columns

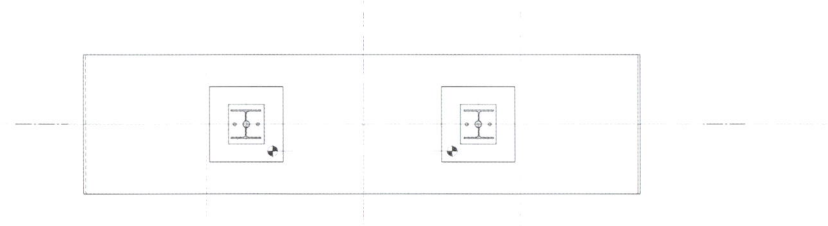

⑩ The walk provides a good place for viewing Hamburg's skyline
⑪–⑬ The walk offers views in every direction

Hovenring bicycle bridge

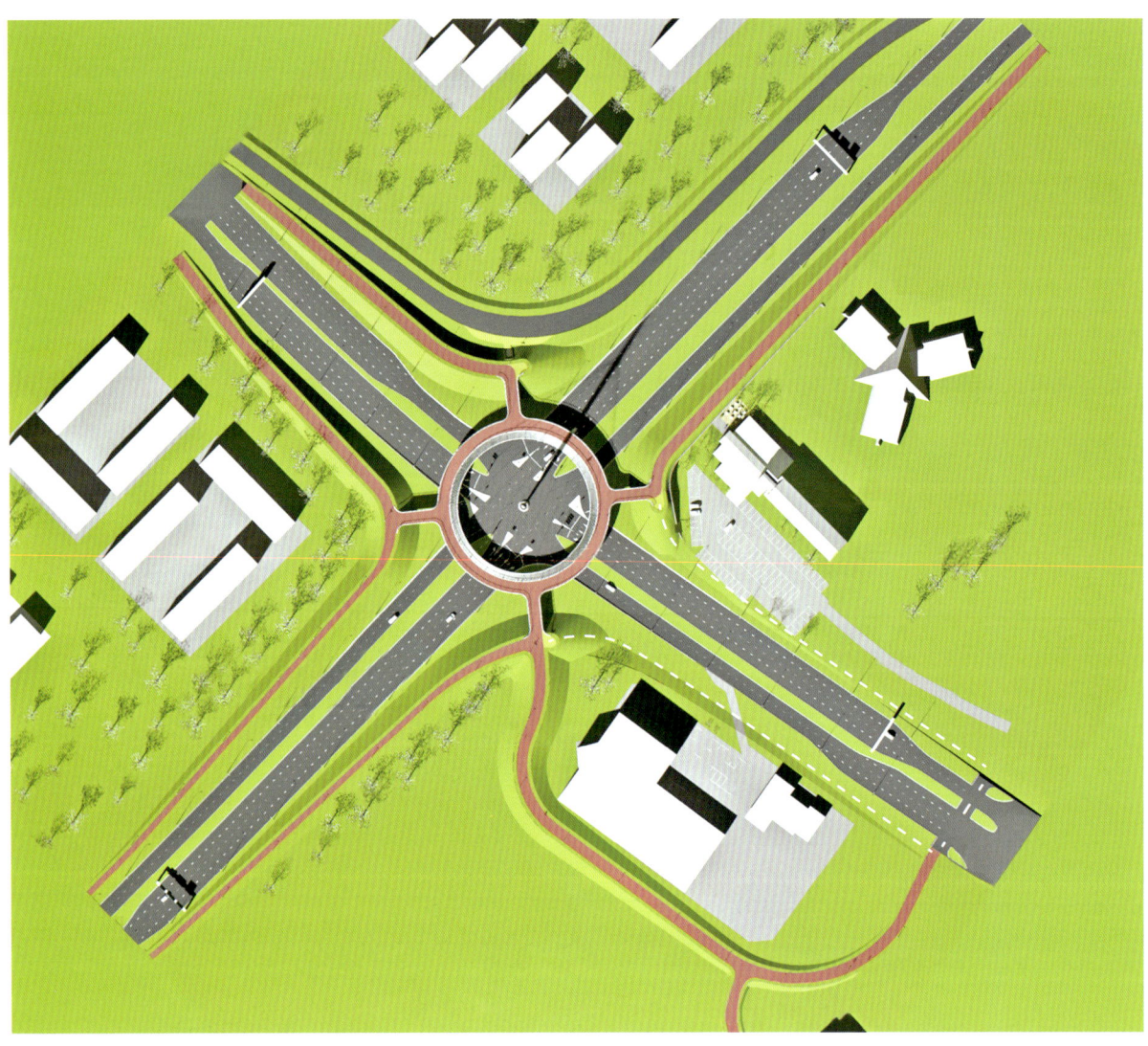

Site plan

● Location / Eindhoven, the Netherlands ● Length / 984 feet (300 meters) ● Completion / 2012 ● Landscape design / ipv Delft ● Photographer / HenkSnaterse/ipv Delft ● Client / Eindhoven City Council and Environment ● Budget / €6,300,000

The development of a residential area near the Heerbaan/Meerenakkerweg intersection meant the previous level crossing needed changing in order to cope with the growing traffic. As Eindhoven City Council refrained from using cyclist underpasses and didn't want a level crossing roundabout either, a circular cable-stayed bridge appeared to be the best option. This cable-stayed circular bridge offers cyclists and pedestrians an exciting crossing point. Like a flying saucer, the steel bridge hovers above the Heerbaan/Meerenakkerweg intersection. Its impressive pylon marks the entrance way to the cities of Eindhoven and Veldhoven.

The steel bridge comprises a 229.66-foot-high (70-meter-high) pylon, 24 steel cables and a circular bridge deck. The cables are attached to the inner side of the bridge deck, where the bridge deck connects to the circular counter weight, so that the torsion within the bridge deck is prevented. The M-shaped supports near the approach spans also ensure stability. One of the challenges in the design process was the spatial integration. The existing infrastructure and buildings set the boundaries for the grades of the slopes leading up to the roundabout. As space was limited, it was decided that the ground level of the intersection underneath should be lowered by 4.92 feet (1.5 meters), allowing for comfortable slopes for pedestrians and cyclists. The design and building process met many technical challenges. The designer had a very clear view of what the bridge should look like: little more than a thin, circular bridge deck and a powerfully shaped pylon. This meant the standard way of attaching cables to the pylon wouldn't suffice, as it would result in a bulk of steel near the pylon top. Therefore, a tailor-made solution was designed. The same applies to the M-shaped supports near the span bridges.

Befitting Eindhoven's identity as the "City of Light" (Eindhoven is home to the Philips company), the designer also made a lighting design for the Hovenring. One of its main elements is integrated into the circular deck. The space between the counter weight and the deck has been fitted with aluminum lamellas, translucent sheeting, and tube lighting, which at night results in a clearly visible ring of light. Together with the illuminated pylon, the ring of light ensures the bridge's spectacular appearance at night. The functional lighting is integrated into the railing, where LED-lighting illuminates the bridge deck, and ensures facial recognition of the bridge users at the same time. Lights attached to a cable framework between pylon and bridge deck and to the inner surface of the circular counterweight illuminate the intersection underneath. Both day and night, the pylon bridge is a true eye-catcher, spectacular in its simplicity.

01 Bird's-eye view of the Hovenring

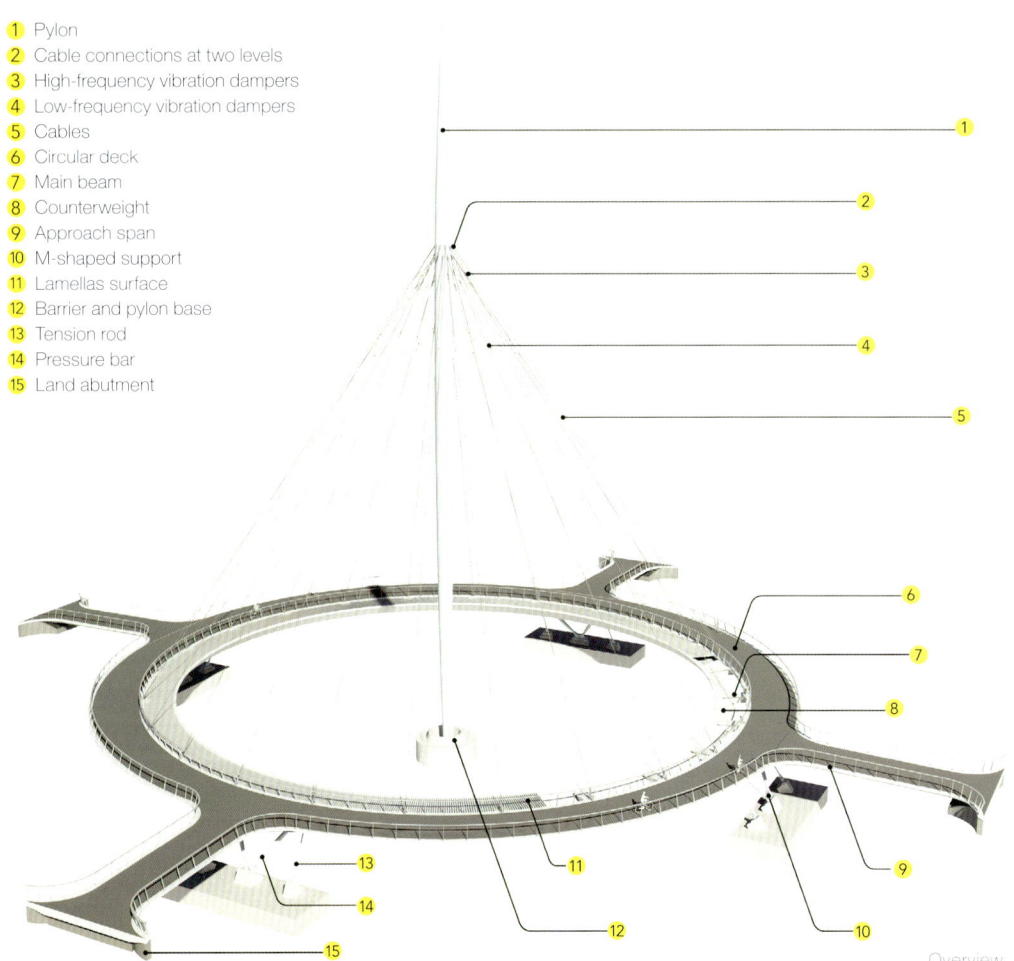

1. Pylon
2. Cable connections at two levels
3. High-frequency vibration dampers
4. Low-frequency vibration dampers
5. Cables
6. Circular deck
7. Main beam
8. Counterweight
9. Approach span
10. M-shaped support
11. Lamellas surface
12. Barrier and pylon base
13. Tension rod
14. Pressure bar
15. Land abutment

(02) The Hovenring is the entrance and a landmark for the city of Eindhoven
(03) The cables are attached to the inner side of the bridge deck, right where the bridge deck connects to the circular counterweight
(04) The M-shaped supports ensure stability

05 – 06 Comfortable slopes for cyclists and pedestrians
07 The ring of light ensures the bridge's spectacular appearance at night
08 Ring of light as seen by traffic underneath

Section view

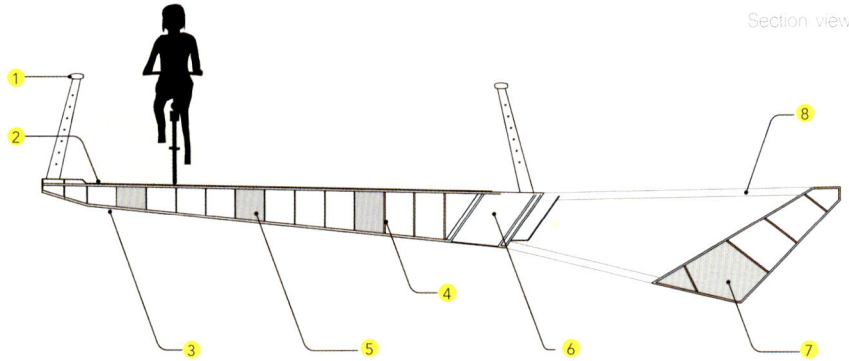

1 Rail with lighting
2 Deck plate with wearing surface
3 Bottom plate
4 Ribs
5 Concete filling
6 Cable connection integrated into deck
7 Counterweight
8 Lamellas surface with lighting

07

08

The Paleisbrug

01 The bridge forms a connection between the old city center and the new Palace District of 's-Hertogenbosch

02 View from the old city center—a wonderful addition to the city of 's-Hertogenbosch

● Location / 's-Hertogenbosch, Netherlands ● Area / 26,909 square feet (2500 square meters) ● Completion / 2015 ● Landscape design / Benthem Crouwel Architects, Piet Oudolf ● Photographer / Jannes Linders ● Client / City of 's-Hertogenbosch ● Budget / €16,000,000

Paleisbrug (Palace Bridge) is a raised park, as well as a pedestrian and cyclist bridge in one. The bridge forms a 820.21-foot (250-meter) link across the railway track between the old town center and the new Paleiskwartier (palace district) of the city of 's-Hertogenbosch, creating a sense of unity with the surrounding cultural landscape. From Paleisbrug, there is a magnificent view over the Gement, the grassland that was used as an inundation area during the Eighty Years' War (1568–1648).

Plants, trees, benches, and lighting have been integrated in folded sheets of weather-proof steel. The rusty color of this steel fits in well with the atmosphere created by the city's fortifications. Weather-proof steel is a steel alloy with a dense corrosion layer, which limits corrosion and means the metal can be left exposed. The bridge has a lifespan of at least 100 years.

The Paleisbrug is composed of various sizes of spans that run from column to column. The columns are covered in weather-proof steel panels, so that they form a whole with the spans. The spans comprise horizontal weather-proof steel ducts covering a concrete compression layer. The longer the span is, the higher the ducts are. The largest span, measuring 196.85 feet (60 meters) in length, runs across the railway track. For this span, an additional trussed joint has been created on top of the bridge deck. A structure underneath the bridge deck was not possible here. The trains need such a large amount of headroom that the entire bridge would have to be raised by a couple of meters. The concrete compression layer is perforated in areas with trees. This has created a tree bed in the steel box girders. The perforations are placed strategically and lead to a good result, both for the planting scheme and the structure. The concrete is not visible anywhere on the bridge. Four lanes, each 6.56 feet (2 meters) wide and alternately consisting of flower beds and paving,

Master plan

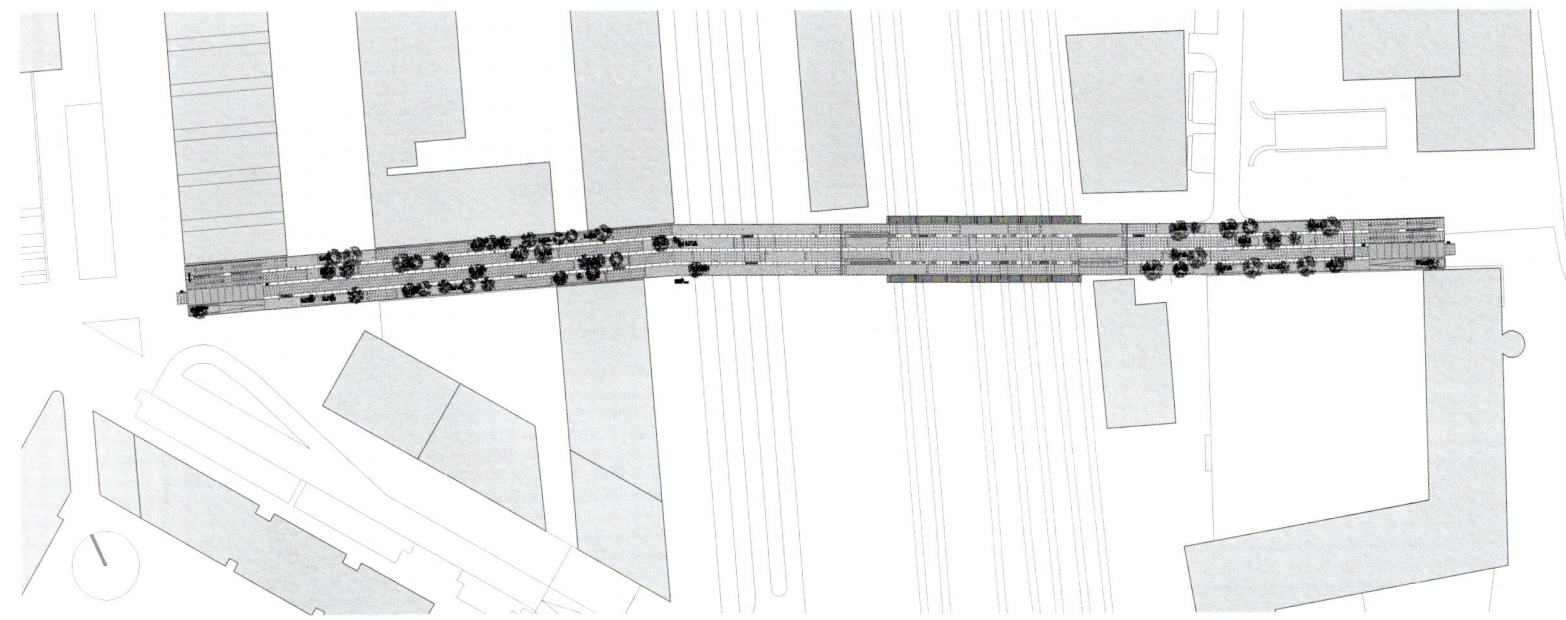

run across the length of the bridge. Strips, 23.62 inches (60 centimeters) wide and finished with a weatherproof steel-studded sheet, run between the flower beds and paving. This is where the cables, pipes, and gutters for the rainwater discharge from the street can be found. The setup of the bridge deck and pipework can be replaced, if necessary, without affecting the main supporting structure.

The plants and shrubs on the bridge are divided into three zones, each with its own character. A savannah-like vegetation is interspersed with high vegetation and trees. In the middle of the bridge, the predominantly low vegetation echoes the grand view. The plants are selected for different flowering periods, making the bridge special throughout the year. A drip-feed watering system with detection has been installed in the flower beds on the bridge. In the evenings, the plants, benches, and paths are lit up by LED lighting, which means the bridge is a pleasant area even after sunset.

Floor heating provides a low temperature (50 °F/10 °C) on the deck and the stairs of the bridge. This prevents the bridge from freezing in the winter without needing to spread salt, which would affect the steel and the plants. The bridge will also be used as a massive solar collector during summer, since the floor heating is connected to an Aquifer Thermal Energy Storage (ATES) system that releases the stored heat during winter.

Paleisbrug has incorporated novel sustainable techniques and ecology-friendly concepts into the design for pedestrians and cyclists. It serves as a model for future landscape architecture.

(03) The bridge has created a lively space embraced by all inhabitants from the moment of completion
(04) It is easily accessible for all kinds of users because of the integrated escalator

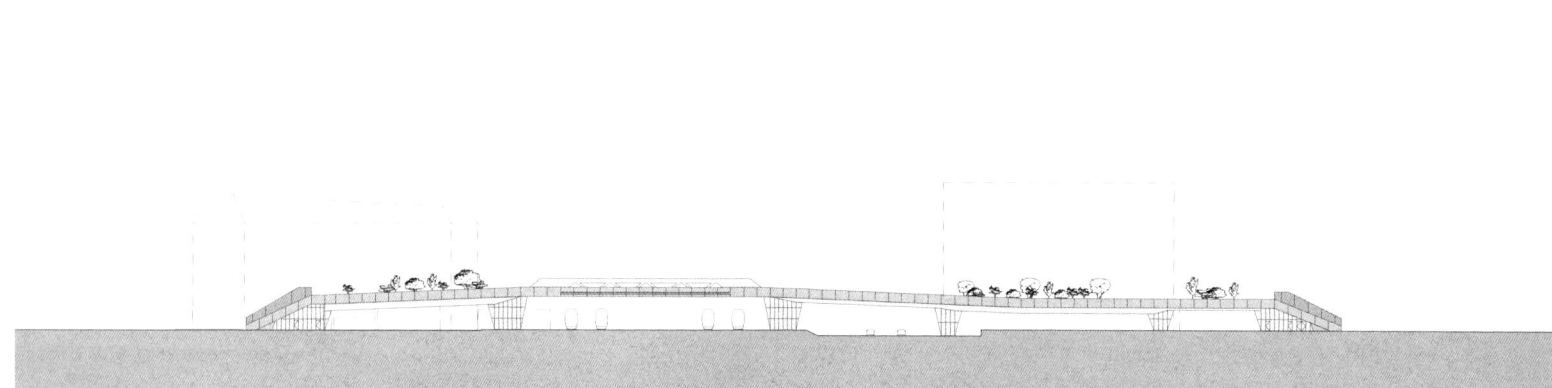
Elevation

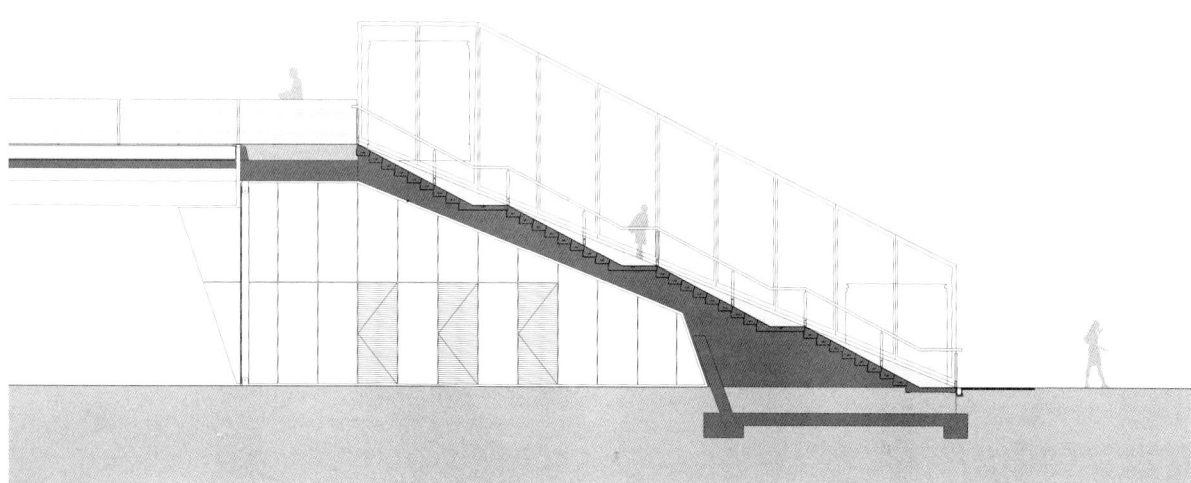
Stairs and elevator

243

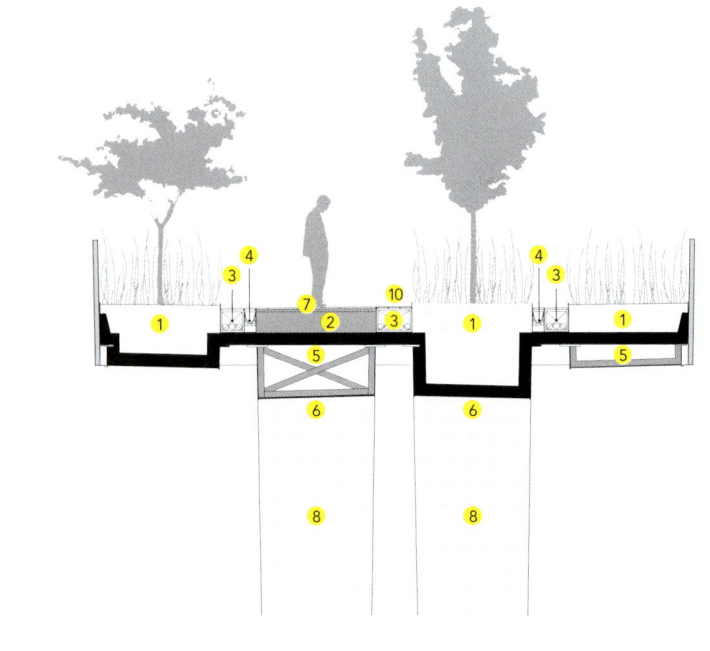

Section 1

1	Planters
2	Walkway
3	Gutter for cables
4	Gutter
5	Concrete
6	Weather-proof steel
7	Tiling with floor heating
8	Concrete columns with weather-proof steel
9	Truss
10	Studded sheet of weather-proof steel

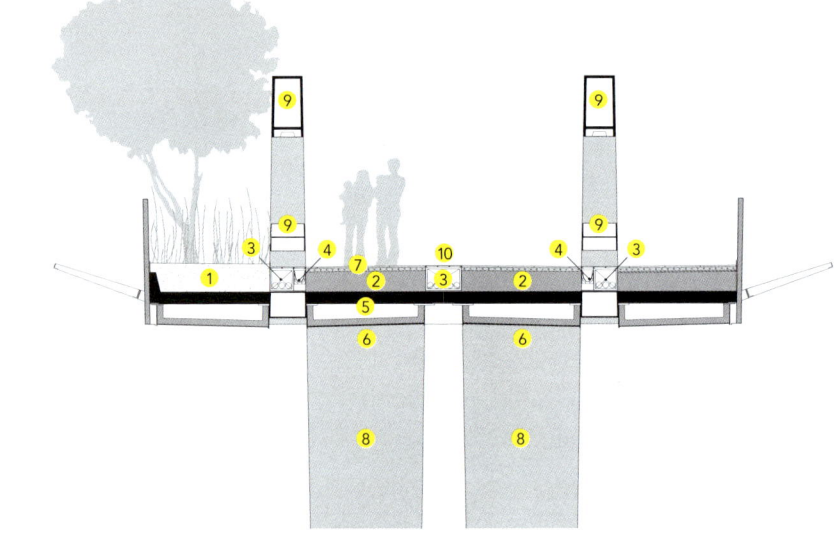

Section 2

05 Its glass escalator is unique
06 The bridge has a meandering path of plants and benches
07 The bridge invites visitors to stop for a while with its inviting benches and good lighting

06

07

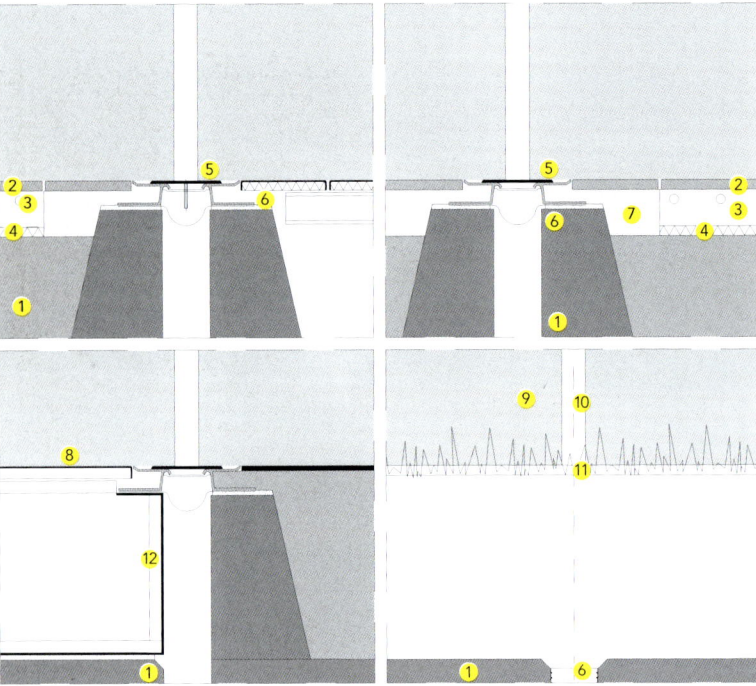

1. Concrete
2. Ceramic tiles
3. Ceramic screed floor
4. Synthetic floor with heating
5. Stainless-steel strip
6. Dilation section
7. Shrink grout
8. Weathering steel-studded panels
9. Elevation parapet
10. Joint of parapet
11. Plant compartments
12. Channel element

Detail 1

Fragment

08 The Paleisbrug forms a unique view, looking over the rail tracks and city from both sides of the bridge
09 The design is robust, fitting with the industrial surroundings
10 It has become a favorite spot for many residents of 's-Hertogenbosch

Vlotwateringbrug

Aerial plan

- Location / Monster, The Netherlands ● Length / 82 feet (25 meters)
- Completion / 2015 ● Landscape design / NEXT Architects
- Photographer / Raymond Rutting, NEXT Architects ● Client / Municipality of Wateringen ● Budget / €700,000

The Vlotwateringbrug, or popularly known as the "bat bridge," is a unique bridge. It was built not only to provide a pedestrian and bicycle connection to Peolzone, but also to function as an ideal habitat for accommodating various species of bats. It is discussed as a textbook example of how a functional object can, at the same time, serve nature.

Crossing the Vlotwatering, the new bridge winds between two existing parcels and marks the entrance of the Poelzone. At the highest point, the bridge makes a turn for a clear view of the area. The wooden lamellas provide openness to the natural banks along the water. With a length of 82.02 feet (25 meters), the concrete arch spans the entire Vlotwatering. The S-shaped deck is supported with a pressure arc that slants under the bridge.

The project is located along a flight route of several bat species. To optimize the suitability of the bridge for bats, the structure is made out of concrete. The mass of the concrete provides a stable and pleasant climate. The underside of the bridge is provided with entrance slots, part of a pattern of grooves in the concrete arch. The openings have a rough finish for grip. Clever use is made of the structural space in the cross-section to be used as roosts.

The bat bridge has become part of the Poelzone. It is not only a pedestrian and bicycle bridge, but also an elongated area in the Westland between the existing centers of 's-Gravenzande, Naaldwijk, and Monster. Along with the waterway, there is a new green recreational zone with natural banks and spawn sites for fish. The landscape design is made to strengthen the existing green and ecological connections in which natural and recreational values complement each other.

Section backbone of the bridge, vertical slats and winter accommodations in bridge head

Section bridge head, winter accommodation and one of the entrances to the bridge

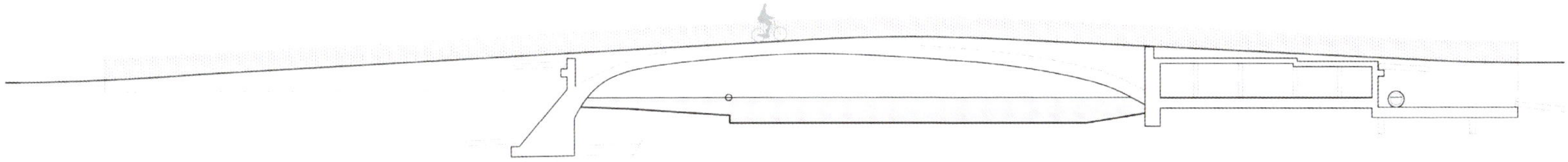

Cross-section of bridge and vertical slats

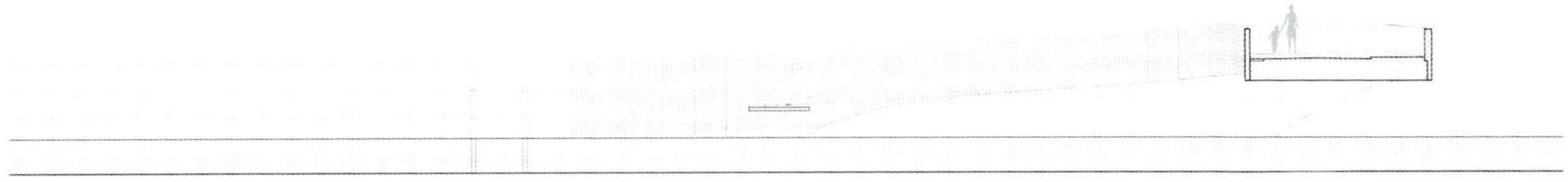

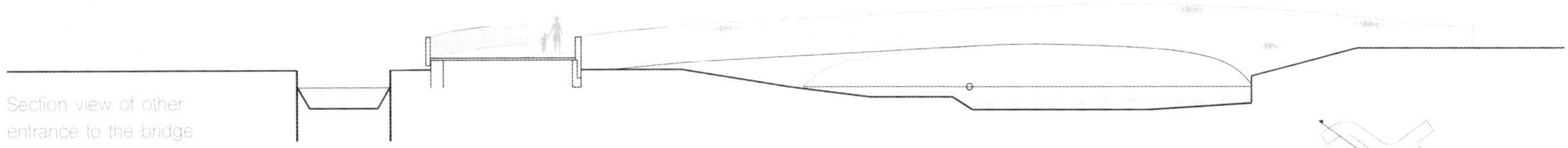

Section view of other entrance to the bridge

(01) Entrance of winter accommodation for bats

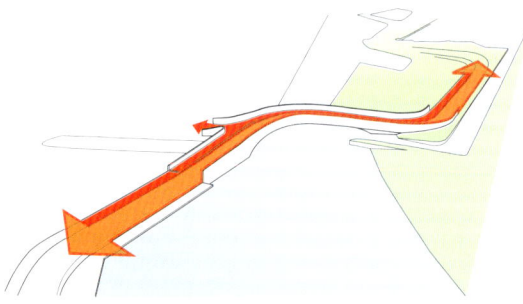
Continuity route

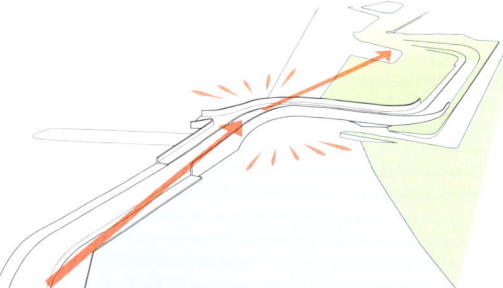
Bridge as entrance of the route

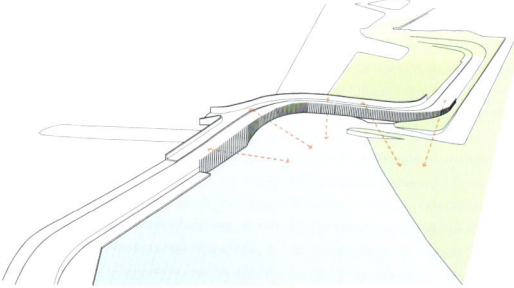
Asymmetry for targeted openness

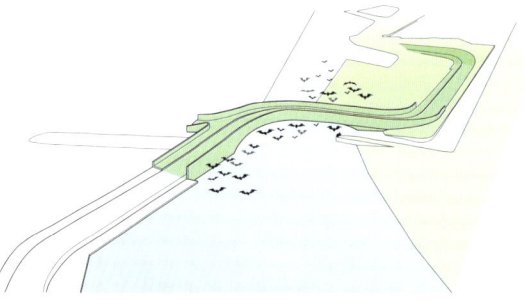
The bridge functions as accommodation for bats, as part of the main route

02 View of the bicycle bridge
03 View of bridge melting into existing surroundings
04 Access gaps for bats' summer accommodations in the concrete mass under the bridge

05

06

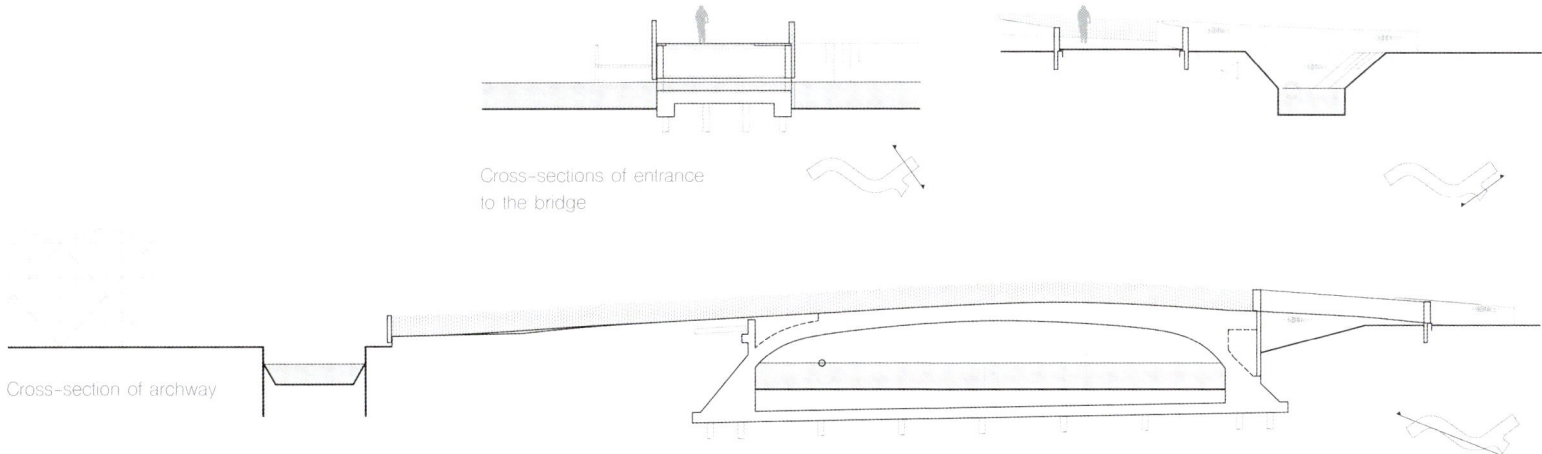

Cross-sections of entrance to the bridge

Cross-section of archway

- 05 Asymmetric bricks and timber slats
- 06 Entrance from parking lot
- 07 Pedestrian zone next to bicycle path
- 08 African padauk timber slats

La Sallaz Footbridge

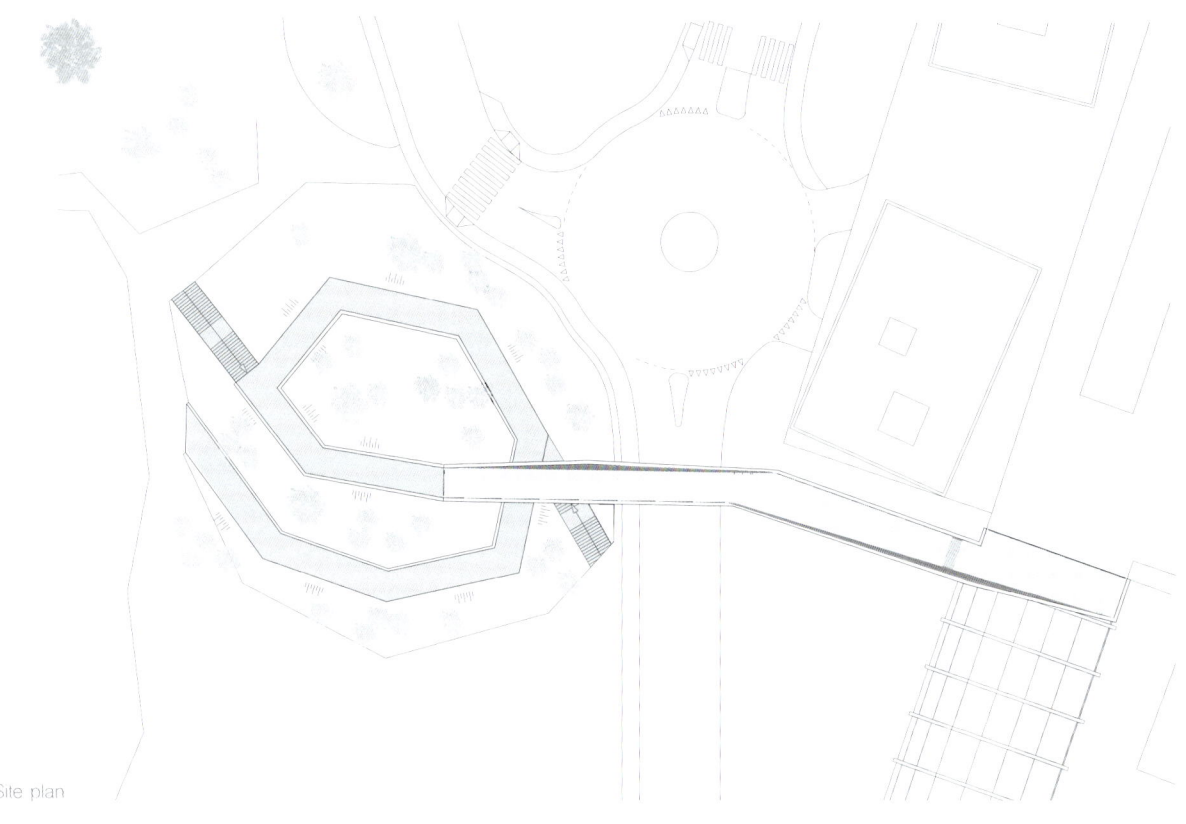

Site plan

- Location / **Lausanne, Switzerland** ● Length / **233 feet (71 meters)** ● Completion / **2012** ● Architect / **2b architects** ● Landscape design / **Cécile Albana Presset** ● Photographer / **Roger Frei** ● Client / **City of Lausanne** ● Budget / **Confidential**

La Sallaz Footbridge is a pedestrian extension between the Metro station and the Bois de Sauvabelin. The design team's goal was to create a connection between the urban surroundings and the artificial natural landscape of Le Vallon.

The duality of different geometries at the location and the quadrangular construction on the square, which is compared to the so-called "natural" character of Le Vallon, became a keynote for the bridge construction. Depending on perception, that is, whether one crosses over the bridge as a pedestrian, or under it as a car driver, the La Sallaz Footbridge design presents its play of criss-crossing lines between the square and the street, thereby shifting them together into a characteristic form. It is a pleasant and fresh experience to walk on the bridge, but it also becomes an attractive entrance landscape of the city.

To announce its dual role as a new gateway into the city and as a connecting element for slower foot traffic, the design uses the two complementary materials of wood and concrete. The interplay between specific material properties creates contrasting atmospheres in experiences of the bridge. The first is for the drivers progressing along the street below, and the second for the pedestrians crossing the heart of the bridge above the car traffic. On the opposite side, the development of La Sallaz Footbridge rests on the mound of rubble created while the Metro was being constructed. There, the atmospherically charged course is continued over the ramp, leading downwards in a large spiral through the dense ornamental shrubbery. Lights are installed at the foot of the railings throughout the whole bridge, making it a safe pedestrian experience at night and also a beautiful icon for automobiles when they enter the city.

01 View from the road towards the concrete beam and pillars
02 Entrance of the spiral ramp
03 Wood cladding on the interior side of the bridge

04 Concrete footbridge lying on the hill
05 Wood cladding detail
06 Interior view of the pedestrian footbridge looking towards the Sauvabelin forest
07 General night view of the footbridge
08 Lights are installed at the foot of the railings throughout the whole bridge

Detail section

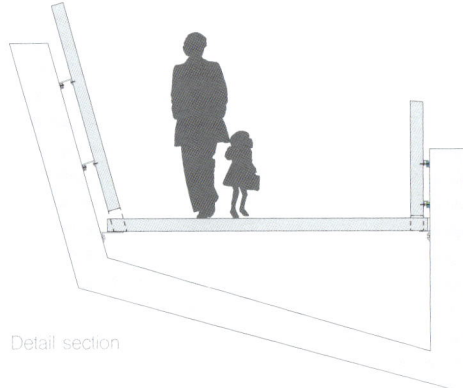

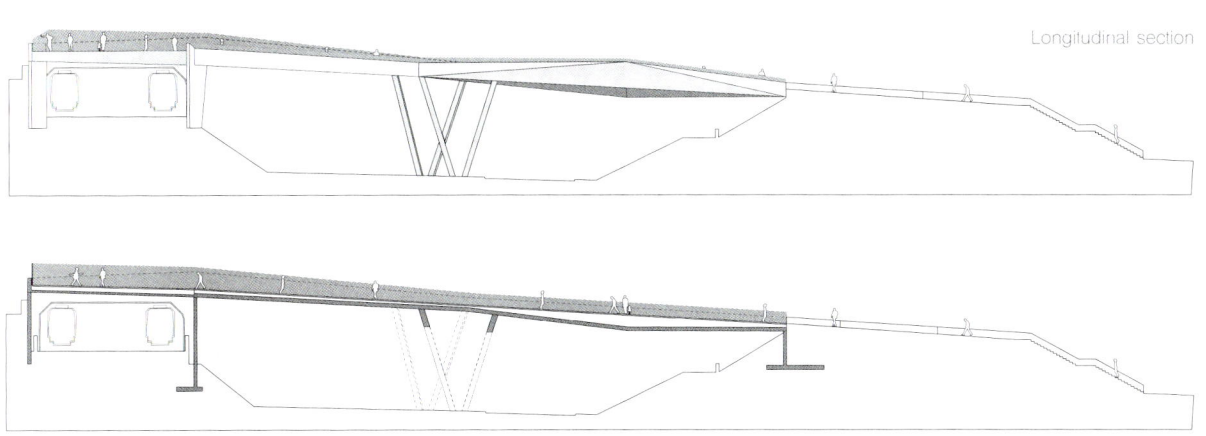

Longitudinal section

Pedestrian Bridge in Aranzadi Park

● Location / **Pamplona, Spain** ● Area / **6.79 acres (2.74 hectares)**
● Completion / **2015** ● Landscape design / **Peralta Ayesa Architectos** ● Photographer / **Pedro Pegenaute, Eduardo Berián/ Hidrone** ● Client / **Pamplona City Council** ● Budget / **€92,993,174**

This project aims to improve access between the park and the city. In the park system, the bridge serves not only as a separate object, but also forms a part of the geometric design of the pedestrian and bicycle path. Joining the river system, the path travels from Errotazar Street, thus avoiding the current road traffic, creating an elevated pedestrian passage, and connecting to the Enviromental Education Museum. At the same time, it creates an ideal spot for visitors and an observation point from the bridge to view the park from the city. The landing in the Aranzadi Park, located at the highest point of the Plaza de los Manzanos, next to the banks of the Arga River, is a gathering point for the pedestrians and cyclists traveling across the perimeter on their way to the plaza.

An L-shaped layout has been proposed, with two orthogonal and opposing branches. Each branch has a unique antagonistic function, based on its form. There is structural purity and clarity in both elements. Its layout, parallel to the river, located partially in the floodable area, is stony, pinned, and anchored to the ground. Its slope has an angle of less than 6 percent, rising in the direction of the current, without offering resistance to future avenues, and resembling the bridges, ramps, and concrete paths of the park. The perpendicular and horizontal design is light, svelte, and aerial, suggesting that it is a bridge suspended between two realities. The perforated, weathered steel is located in both the parapets and the pavement, reinforcing its buoyancy and transparency while blending in with the natural environment.

Artificial lighting reinforces the essence of each part. The concrete element is illuminated along the line of the park, thanks to the suspended projectors that offer uniform, superficial lighting. The weathered steel panel hides a double line of warmer lighting, which allows for a volumetric illumination of the panel, backlighting the perforated surface, and emphasizing its warm and volatile nature.

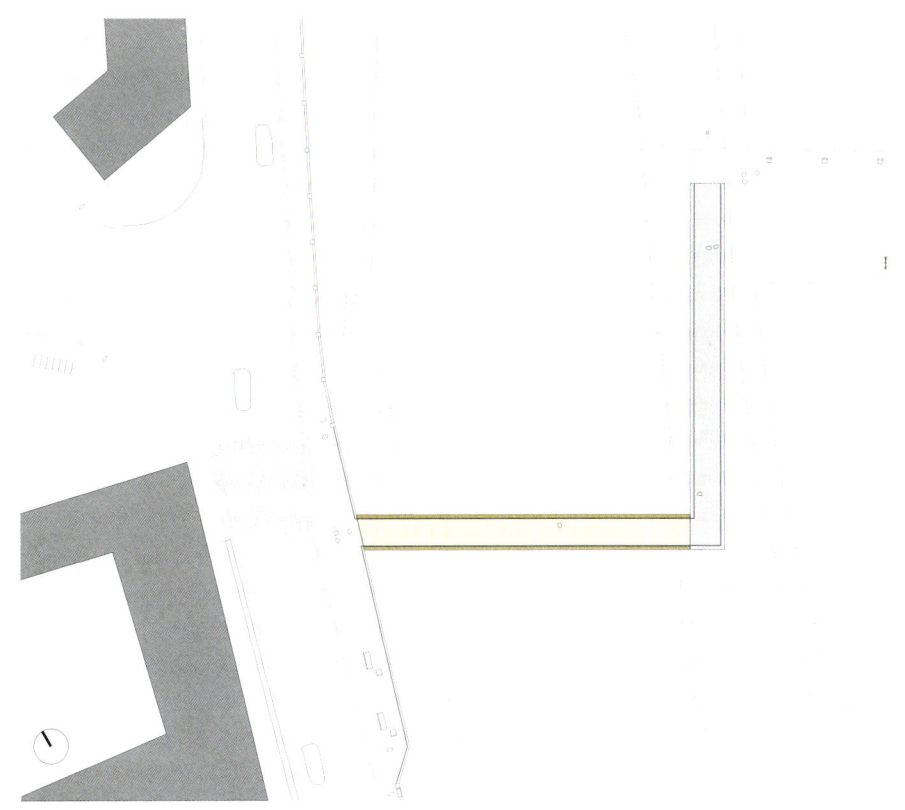

General floor

01. Aerial view from Aranzadi Park
02. Concrete access ramp in the foreground and the weathered steel panel at the bottom

Park elevation

General cross-section

03

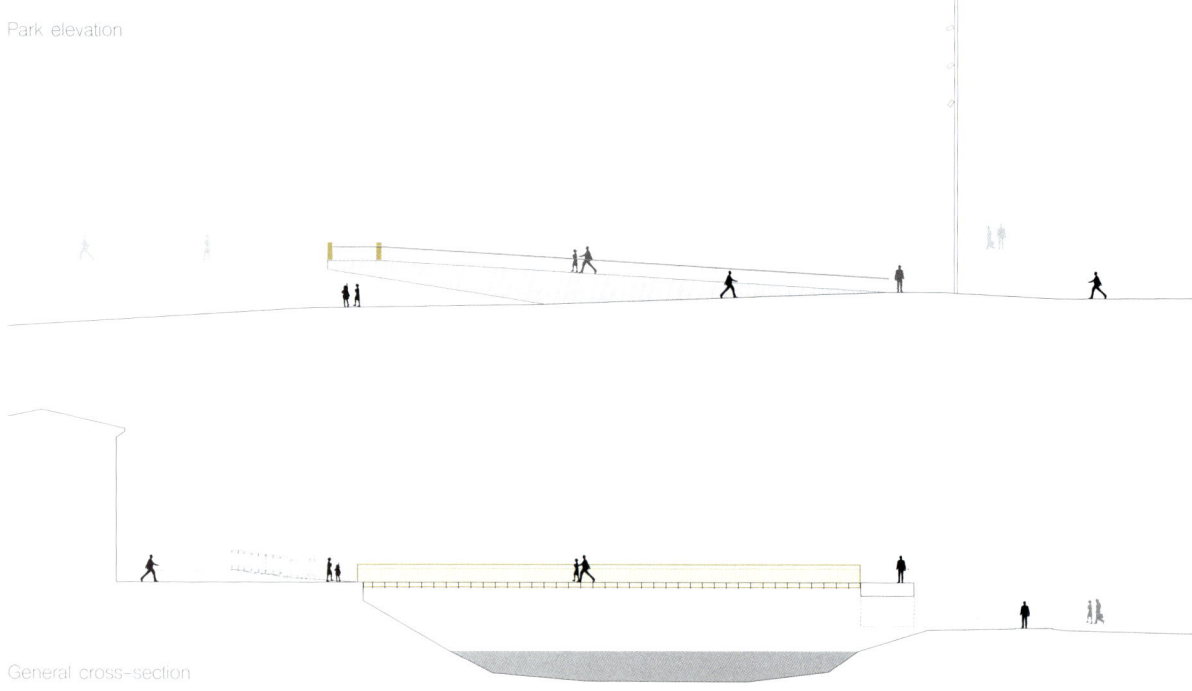

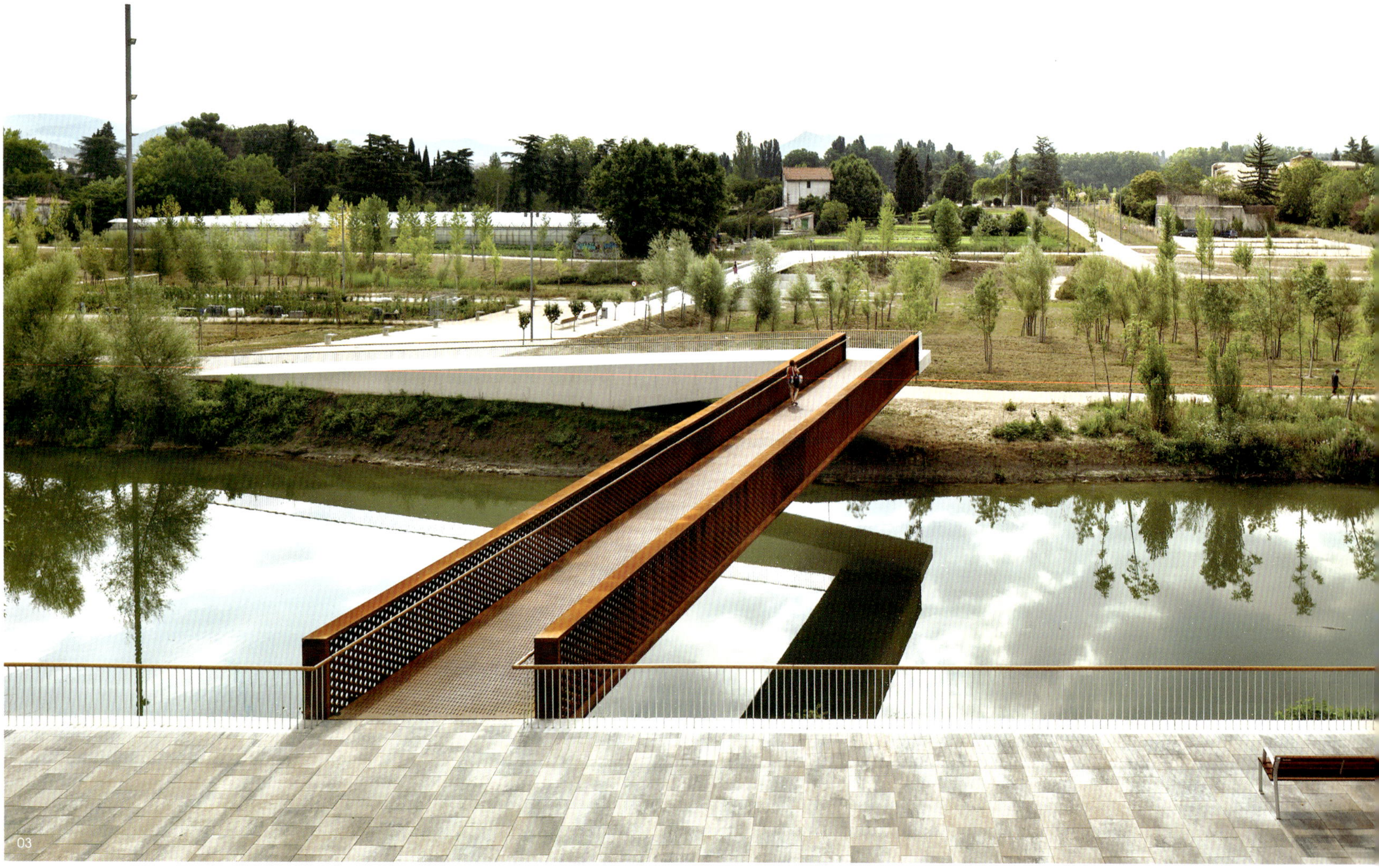

(03) Aerial view from the Environmental Education Museum
(04) The weathered steel panel floating on the Arga River
(05) View from the Aranzadi Park

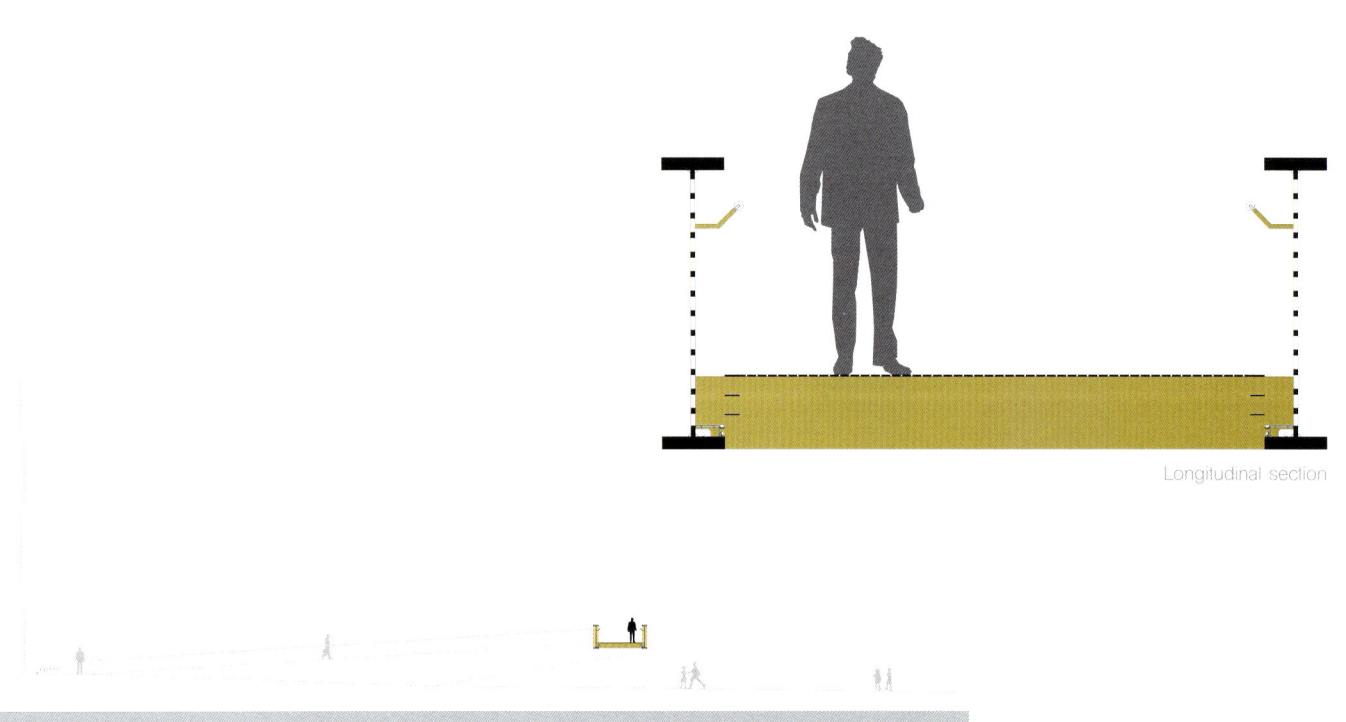

Longitudinal section

Section of the pedestrian bridge

06 Illumination of the panel
07 The perforated, weathered steel is located in both parapets and pavement, reinforcing its buoyancy and transparency
08 The panel hides a double line of warmer lighting, backlighting the perforated surface

The Bicycle Snake

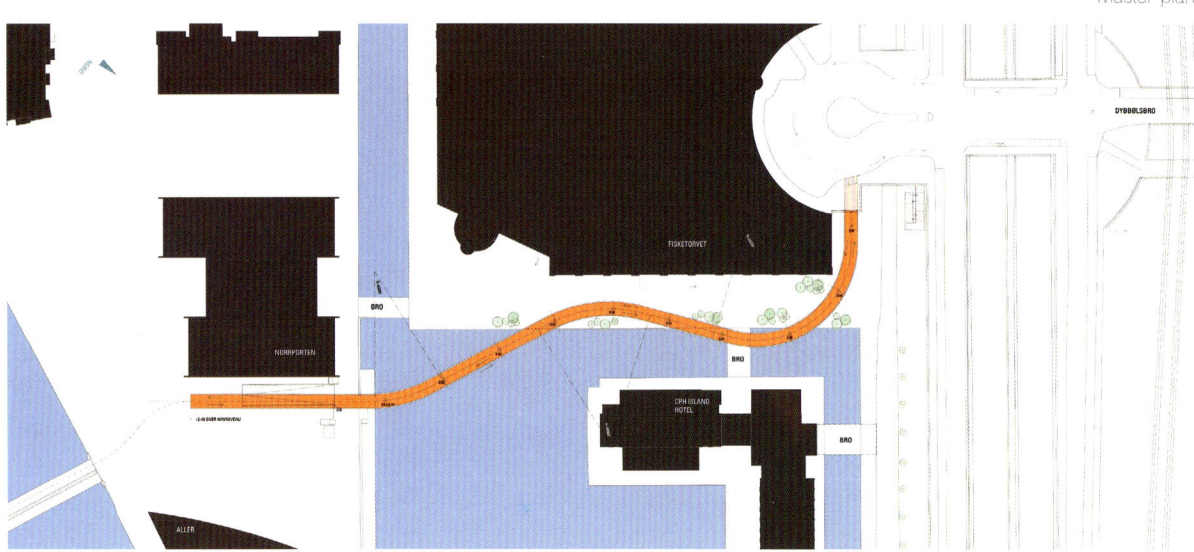

Master plan

● Location / **Copenhagen, Denmark** ● Length / **754.6 feet (230 meters)** ● Completion / **2014** ● Architect / **DISSING+WEITLING architecture** ● Landscape architect / **Marianne Levinsen Landskab** ● Photographer / **Ole Malling, Rasmus Hjortshøj/COAST Studio, DISSING + WEITLING Architecture** ● Client / **Municipality of Copenhagen** ● Budget / **DKK38,000,000**

The City of Copenhagen has published its 2011–15 planning strategy for becoming the world's best bicycle town to promote a healthier lifestyle, and set the goal to become a CO_2-neutral city by 2025. The Bicycle Snake (or Cykelslangen) is a project among a series of initiatives to fulfil this strategy. With the change from commercial harbor activities to residences and retail, the inner harbor of Copenhagen has undergone a pronounced transformation. As part of this transformation, the first pedestrian and bicycle crossing of the harbor's Brygge Bridge was built to connect the two parts of the city. However, heading to or from Brygge Bridge on the eastern side of the harbor, cyclists had to carry their bikes down or up a full flight of stairs at one end of the quayside. Thus, the Bicycle Snake, a 754.59-foot-long (230-meter-long) sky bridge, was built as a short-cut to Brygge Bridge. It takes off where Brygge Bridge ends on the eastern side of the harbor and continues in a meandering course to Kalvebod Brygge (a major roadway), some 18.04 feet (5.5 meters) above the quay.

The client's brief called for a somewhat minimal bicycle ramp, which could provide an alternative to the staircase. The architects saw tremendous potential for the new ramp to become something more than just a replacement for the staircase. Through unfolding the ramp from the corner site of the staircase, stretching it out and curving it across the water, in between the buildings, and down close to the Brygge Bridge, a clear pathway was sketched out. The design makes it more joyful to ride on the bridge, with shallower gradients and better curvature, and pulls together an area with a multitude of disparate buildings. This could also make cyclists inadvertently slow down, creating a safe cycling environment.

The architects strived for a slim structure, with all parts being structural, so as to reduce the visual impact. The bridge is a painted, airtight, welded-steel structure, carried by a central steel spine, a 29.53-inch (75-centimeter) box girder, from which a series of cantilevered struts, made of folded steel plates, carries the steel plate deck. The parapet, consisting of inward-leaning steel bars, is conceived as a transparent film with no modular hierarchy to underline the fluidity of movement through space and celebrate cycling through looking out and being seen. The design style of the bridge relates essentially to Brygge Bridge, but differs in detail, reflecting its functionality, alignment, and setting. The anti-slip pavement consists of granulated stone on an acrylic compound. The bridge is lit by LED strip lighting built into the parapet handrail. The thin vertical posts are backlit at night, highlighting the winding path.

The Bicycle Snake is used by 12,500 cyclists every day now. The super cycle track generates better conditions for commuter cyclists, and motivates more people to take bikes instead of cars. It epitomizes the image of Copenhagen as a bicycle city.

01 Bird's-eye view of the Bicycle Snake
02 The bridge is about 18 feet (5.5 meters) above the quay
03 The bridge meanders between the buildings like a snake

④-⑤ The bridge completely separates cyclists from pedestrians, and has solved a large logistic problem
⑥ Activities at the harbor front
⑦ Cyclists riding on the bridge

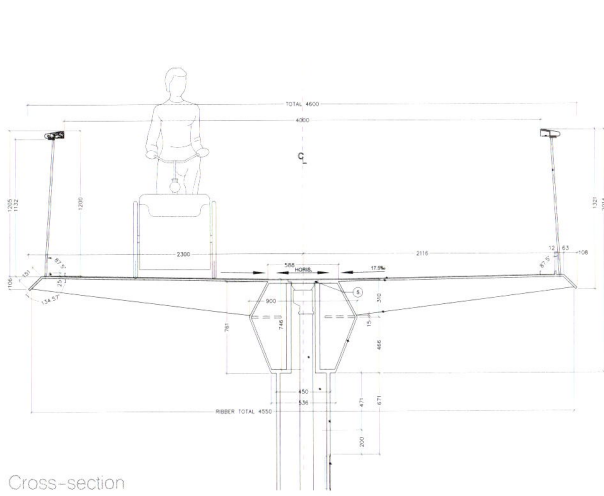

Cross-section

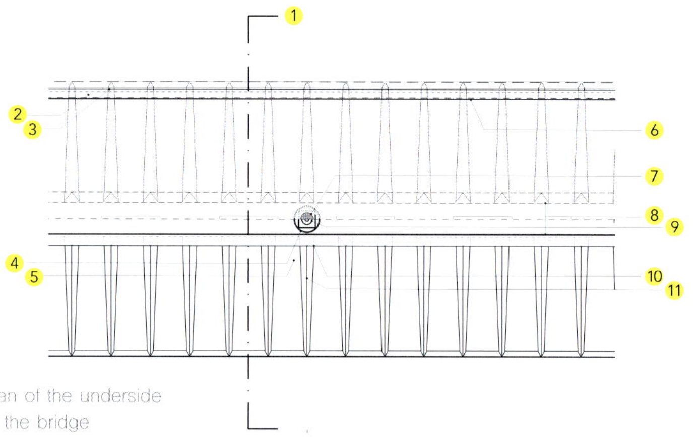

Plan of the underside of the bridge

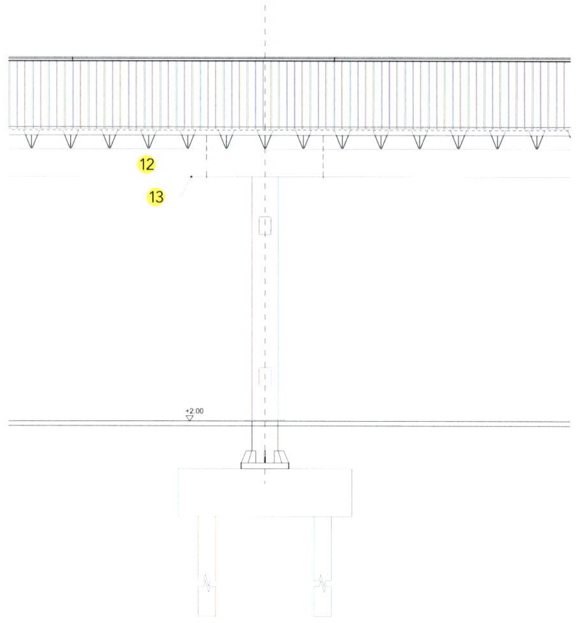

Elevation toward the north-west

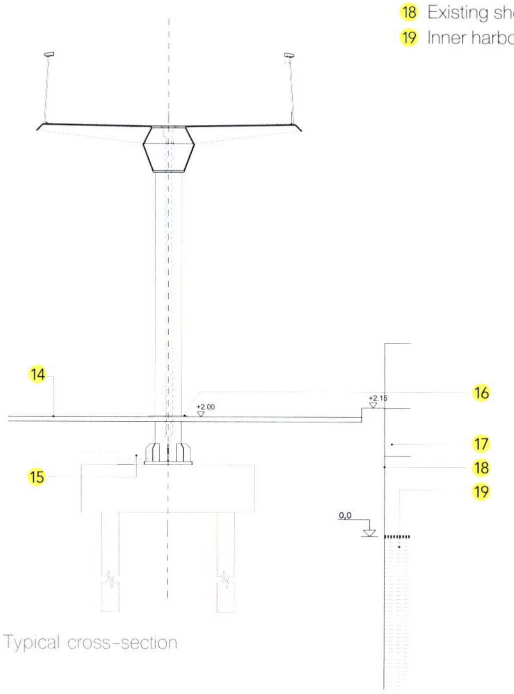

Typical cross-section

1. Typical cross-section
2. Handrailing
3. Baluster
4. Column
5. Underside of the steel deck
6. Light built into handrailing
7. Drainage grid
8. Soffit of bridge deck spine
9. Drainage
10. Edge of diaphragm
11. Cantilever transverse rib
12. Center line of bridge deck
13. Deck spine
14. Typical level of existing quay
15. Drainage to existing sewer
16. Foot plate
17. Existing footbridge
18. Existing sheet pile
19. Inner harbor

08 View of the bridge from below
09 Structure of the underside of the bridge
10 – 11 The parapet is simple and transparent, to underline the fluidity of movement

Index

2b architects
P. 254 Av. de Beaumont 22a, CH–1012 Lausanne, Switzerland
Telephone • +41216175817
Fax • +41216175833
Website • www.2barchitectes.ch

Agraz Arquitectos
P. 120 San Felipe #872 Col. Capilla de Jesús C.P. 44200 Guadalajara, Jalisco, Mexico
Telephone • +523338274500
Website • agrazarquitectos.com

Alta/Greenways, LLC
P. 190 111 East Chapel Hill Street, Suite 100, Durham, NC 27701, United States
Telephone • +319194848448
Website • altaplanning.com

Arplan, Ltd
P. 48 10/12 Meistaru Street, LV-1050, Riga, Latvia
Telephone • +37167227102
Fax • +37167227103
Website • www.arplan.lv

ASPECT Studios
P. 78 Studio 61, Level 6, 61 Marlborough Street, Surry Hills, NSW, 2010, Australia
Telephone • +61296997182
Fax • +61296997192
Website • www.aspect.net.au

Benthem Crouwel Architects
P. 240 Generaal Vetterstraat 61, NL-1059 BT Amsterdam NL, The Netherlands
Telephone • +31206420105
Website • benthemcrouwel.com

BKK Architects
P. 100 Level 9, 180 Russell St, Melbourne 3000, Victoria, Australia
Telephone • +61396714555
Website • www.b-k-k.com.au

BMLA Ltd.
P. 198 Wanganui Avenue, Herne Bay, 1011, Auckland, New Zealand
Telephone • +642142477524
Website • www.bmla.co.nz

CANAPÉ Cantieriaperti – Architecture and Urban design
P. 186 Via Fondobanchetto, 22 _ 44100 Ferrara, Italy
Telephone • +390532760050
Fax • +390532186236 2
Website • www.progettocanape.eu

Corkery Consulting Pty. Ltd.
P. 84 Suite 3, 38 Albany Street, St Leonards NSW 2065, Australia
Telephone • +61299066636
Website • www.corkeryconsulting.com

Cristina Mazzucchelli Green Design
P. 180 Via Valentino Pasini, 420129 Milano, Italy
Telephone • +39335485336
Website • cristinamazzucchelli.com

CZstudio associati
P. 106 Via della Pila, 38, 30175 Porto Marghera (VE), Italy
Telephone • +390410990566
Fax • +390410990565
Website • czstudio.com

DISSING+WEITLING architecture
P. 264 Dronningensgade 68, 1420 København K, Denmak
Telephone • +4532835000
Website • www.dw.dk

Estudio Carme Pinós
P. 174 Avd. Diagonal 490, 3-2 08006 Barcelona, Spain
Telephone • +34934160372
Fax • +34932181473
Website • www.cpinos.com

Group GSA
PP. 126, 132, 146 Level 7, 80 William Street, East Sydney, NSW 2011, Australia
Telephone • +61293614144
Website • www.groupgsa.com

Grupo Aranea Ltd.
P. 214 Avenida General Marvá n°7 3°A/1°B, cp/ 03005, Alicante, Spain
Telephone • +34965921695
Website • grupoaranea.net

häfner jiménez betcke jarosch landschaftsarchitektur gmbh

P. 226 Schwedter Str. 263, 10119 Berlin, Germany
Telephone ● +4903044324160
Fax ● +4903044324159
Website ● www.haefner-jimenez.de

ipv Delft

PP. 210, 234 Oude Delft 39, 2611 BB Delft, The Netherlands
Telephone ● +310157502575
Website ● www.ipvdelft.com/www.ipvdelft.nl

LandLAB

P. 78 Level 2, 17 Sale Street, Freemans Bay, Auckland 1010
Telephone ● +6493795805
Website ● www.landlab.co.nz

M42KStudio

P. 56 Santiago, Chile
Telephone ● +56223547710
Website ● www.mapocho42k.cl

NEXT Architects

P. 248 P.van Vlissingenstr 2a, 1096 BK Amsterdam, The Netherlands
Telephone ● +31020463046
Website ● http://www.nextarchitects.com

Peralta Ayesa Architectos

P. 258 Cultura 1 Oficinas Entreplanta, C3010 Barañáin Navarre, Denmark
Telephone ● +45948183715
Website ● www.peraltaayesa.com

scape Landschaftsarchitekten GmbH

P. 166 Friedrichstrasse 115a, 40217 Düsseldorf, Germany
Telephone ● +492113020370
Fax ● +4921130203720
Website ● www.scape-net.de

Stradivarie Architetti Associati

P. 114 Piazza Benco 4-Trieste 34122, Italy
Telephone ● +390400640442
Website ● www.stradivarie.it

Sturgess Architecture

P. 220 200, 724 11 Avenue SW, Calgary AB T2R 0E4, Canada
Telephone ● +4032635700
Fax ● +4032625710
Website ● www.sturgessarchitecture.com

Urban Agency

P. 92 Kirsten Walthers Vej 9, 2500 Valby, Copenhagen, Denmark
Telephone ● +4533245420
Website ● www.urban-agency.com

Urbicus

P. 64 3, Rue Edme Frémy, 78000 Versailles, France
Telephone ● +330139531435
Fax ● +330139494623
Website ● www.urbicus.fr

Valeri.Zoia Architetti Associate

P. 114, 30010 cavallino-treporti, Venice, Italy
Telephone ● +0415370277
Fax ● +0415379812
Website ● www.valerizoia.it

West 8

PP. 140, 152, 160 Schiehaven 13M, 3024 EC Rotterdam, The Netherlands
Telephone ● +310104855801
Website ● west8.com

Wowhaus Architects

P. 72 Address: Bldg. 4, 4th Floor, Bersenevsky per., 5, Moscow 119072, Russia
Telephone ● +74959882094
Website ● en.wowhaus.ru

ZUS

P. 204 Schieblock, Schiekade 189, Unit 303, 3013 BR Rotterdam, The Netherlands
Telephone ● +31102339409
Website ● www.zus.cc

Published in Australia in 2016 by
The Images Publishing Group Pty Ltd
Shanghai Office
ABN 89 059 734 431
6 Bastow Place, Mulgrave, Victoria 3170, Australia
Tel: +61 3 9561 5544 Fax: +61 3 9561 4860
books@imagespublishing.com
www.imagespublishing.com

Copyright © The Images Publishing Group Pty Ltd 2016
The Images Publishing Group Reference Number: 1265

All rights reserved. Apart from any fair dealing for the purposes of private study, research, criticism or review as permitted under the Copyright Act, no part of this publication may be reproduced, stored in a retrieval system or transmitted in any form by any means, electronic, mechanical, photocopying, recording or otherwise, without the written permission of the publisher.

Title: Paths, Tracks, and Trails: Designing for Pedestrians and Cyclists
Author: Paolo Ceccon and Laura Zampieri
ISBN: 9781864706567

For Catalogue-in-Publication data, please see the National Library of Australia entry

Printed by Everbest Printing (Guangzhou) Co Ltd., in China

IMAGES has included on its website a page for special notices in relation to this and our other publications. Please visit www.imagespublishing.com.

Every effort has been made to trace the original source of copyright material contained in this book. The publishers would be pleased to hear from copyright holders to rectify any errors or omissions. The information and illustrations in this publication have been prepared and supplied by the authors and the contributors. While all reasonable efforts have been made to ensure accuracy, the publishers do not, under any circumstances, accept responsibility for errors, omissions and representations, express or implied.